THE OLD ENGLISH FARMING BOOKS
VOL V. 1861 – 1900

THE OLD ENGLISH FARMING BOOKS

VOL V. 1861 – 1900

G. E. FUSSELL

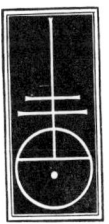

The Pindar Press

London 1991

Copyright © G. E. Fussell 1991

Published by The Pindar Press
66 Lyncroft Gardens
London NW6 1JY

British Library Cataloguing in Publication Data

Fussell, G. E. (George Edwin)
 The old English farming books.
 Vol. 5, 1861–1900
 1. Great Britain. Agriculture, history
 I. Title
 016.630941

ISBN 0 907132 24 3

Printed in Great Britain by
Billing & Sons Ltd, Worcester

CONTENTS

I.	INTRODUCTION	1
II.	1861–1870	2
III.	1870–1900	37
	BIBLIOGRAPHY	131
	INDEX	135

I

Introduction

There is one thing amongst many others that a bibliographer cannot do. I may have been asked it before. The lives of writers of any kind — not only my agricultural pundits — cannot be written, combined and confined within decades: a glimpse of the obvious; this is rather a habit with me. All the same, decades are convenient periods of time within which to confine oneself and this is what I have done here; but I shall open with the details of some books not published strictly within the prescribed limits of this or those of previous volumes of this work. Here I take a liberty overpassing the titular and specified chronological limits.

I must apologise to any readers of my works for any mistakes in dating or editions that I may have made, pleading only the frailty of human nature in which we are all, writers and readers alike, indirectly committed. Such we are and must be.

So, the froth being out of the bottle as Meredith put it in *Diana of the Crossways* we will proceed to the more solid, or in the metaphor, liquid draught in the hope that it will prove both pleasing, palatable and indeed yet more hopefully profitable.

II

1861-1870

Here I must include two writers who were omitted from the previous volume although details of their works really belong there. The first of them is James Main, A.L.S., whose earliest work seems to have been *Illustrations of vegetable physiology practically applied to the cultivation of the garden, the field, the forest containing original observations collected during the experience of fifty years* an 8vo put out by W. Orr, London in 1833. The preface is not modest but what preface is? It expands upon the subject in the words "Some most distinguished naturalists in Europe still but imperfectly understood". The subject had been too frequently explained by reference to animal anatomy, but as mere fictitious knowledge cannot be useful, it is necessary, especially for the young practitioner, that the science should be divested if possible of some of its superfluous disguise and treated of in such language as shall be equal to every capacity. The book was intended to be properly scientific in character. It set out the application of physiological knowledge in sowing, transplanting, propagation, pruning, training etc. Vegetable food must be ascertained by careful analysis and it might safely be inferred that similar bodies or qualities must necessarily be their food. I wonder if this is a cogent argument and whether the conclusion is the proper result of the premises?

This was followed by *The young farmer's manual showing the principles and practices of agriculture to turnip land farms in the South of England with collateral observations and remarks on agricultural cattle, plants, implements etc* a Ridgway 8vo of 1839. Main had no great opinion of the contemporary textbooks. Agricultural writings already before the public, he wrote, are either small practical tracts or voluminous works (I couldn't agree with him more!) The first were too concise, the last far too bulky and diffuse but embrace everything belonging to main chance farming. His own work was of course very different. The following pages, he said in the preface, contain a detail of the actual proceedings and practices of a working farmer who took an active and laborious post in every operation which he describes, though he was restricted by the necessities of a turnip land farm in one of the southern counties, but his lucubrations would, he professed, be useful everywhere. It was originally written as a letter of instruction for a young friend who was about to enter into the business. The book is written in a simple style and is quite exhaustive.

Before this Main had put out *The Villa and Cottage Florists' Directory: being a familiar guide to horticulture* a 12vo London 1830 on which I do not propose to comment. Possibly I am not competent to do so.

Another effort to achieve fame or perhaps be instructive was *The Forest Planters and Pruners Assistant being a practical treatise on the management of the native and exotic forest trees commonly cultivated in Great Britain,* an 8vo illustrated by engraved figures issued by Ridgway & J. Madocks of Ruthin in 1847. I cannot comment on this doubtless profound work. Donaldson said that Main had written *Poultry rearing breeding and fattening* a London 8vo but I shall have to accept his word for this because he stated rather cursorily that Main had written *Popular Botany* a London 8vo of 1835. So much perhaps too much for J.Main.

Sewage is something that is recurrent in life and indeed in the literature detailing its disposal and use for purposes of fertilisation. One, or rather I should say another, writer of an exegesis upon this fertile topic was Lucius H. Spooner who produced *Observations on the present decay among the potatoes with particular reference to the extraction of starch*, which is filed in the R.A.S.E. Library but not in the British Library. It was a London 8vo put out in 1845. Spooner was a vindictive and justified critic of the contemporary methods of sewage disposal and of the neglect of this valuable fertiliser by the contemporary authorities and the indifferent farmers. According to him all the rivers of the country from the Thames to the Tyne were completely and effectively polluted by the contempory indifference to the disposal of urban and human wastes. "The utilisation of sewage by conveyance on and through the land is undoubtedly the only way of purification" or so he said and no doubt believed. Other days, other manners!

In 1851 George Allen Dean published a work under the title of *The land steward. A treatise on all the aspects of farm and estate manangement which are the responsibilty of a land agent or factor*, a large illustrated 8vo of some 304 pp. The preface supplies a synopsis of the book's contents and protests his principles of discussion. It opens with advice on the choice of place, geology, drainage and irrigation, and deals carefully with roads, farm buildings, and general management of estates, including timber and plantations, a comprehensive work. This book was issued again in 1872 as *The culture management and improvement of landed estates with numerous illustrations.* Those of grasses and insects taken from nature, a crown 8vo of a long preface of xxvi pp. and 245 of text. The student of nature would find, he said, that She is incessantly working for the benefit of man, a rather dubious pronouncement. He was very opposed to Mechi's idea of breaking up pasture but thought the question might be left to the cultivators of the soil. It is very pleasant, no doubt, he wrote with some acerbity, for amateurs to talk about farming, but what good that did for the agriculturalist he did not know. It would be instructive to expand his views but that is perhaps unnecessary though attractive.

Another land surveyor, one John Ewart, produced a book *A treatise on agricultural buildings* in 1851 put out by Longmans Green, London, and Oliver & Boyd of Edinburgh. It was of foolscap size and contained many plans and diagrams. He protests that good buildings are essential for good farming and advises landowners that such was the best investment on capital *ergo sum* in the provision of good and convenient buildings, well planned and substantial. There is nothing very original about the diagrams (it would have been impossible!) which are more or less conventional in layout but include everything necessary to the dairy, hog sties and so on.

Two other books not noticed in the previous volume were written one by John Grieve the other by James Haywood. Grieve was an engineer and land surveyor who wrote *The Farmers' Assistant* an illustrated 8vo of 110 pp. published by Gladstone and Hunter of Edinburgh in 1852.

Grieve was not modest. What writer is? He indicated his quite definite opinions in his preface when he pronounced that the first part of this work is intended to impart full information and correct views respecting the theory and practice of land drainage, all unnecessary details being omitted but care taken to correct views and lead to perfect practice. The second part of the book might be termed, he said, the *Agriculturalists Calculator.* It supplied tables for all agricultural purposes, draining tiles, bone manure, dung, guano, weight of cattle and hay, etc, etc, all being carefully revised and checked. Drainage was the great leader of improvement and promoter of fertility. Grieve held what he clearly thought a mistaken view, that any burning land in summer needs no drainage and indeed that it would do injury. Drainage was a topic which he exhausted.

The second work was *Letters to Farmers* by James Haywood, an 8vo of 236 pp. dedicated to Viscount Galway and published by Simpkin Marshall of London in 1852. Haywood claimed that these *Letters* were presented for publication at the request of the members of the Blyth Farmers Club. They were the substance of a series of lectures which he had delivered to that society. The British Library copy of this formidable work was still uncut when I examined it a few years ago. Like most people Haywood was certain of his correctitude, saying that Truth especially in the earlier stages of its discovery and progress rarely escapes being assailed by ignorance and prejudice, an opinion to which it would certainly be possible to subscribe. He discussed the chemistry of the life of plants and animals, manures for crops and feed for beasts: a fairly written and exhaustive piece of work.

George Henry Andrews, an agricultural engineer, was no less exhaustive. It was the era of steam cultivation when he wrote *Modern husbandry; a practical and scientific treatise on agriculture* an 8vo provided with a frontispiece and plates, of some 404 pp. put out by N. Cooke of London in 1853. Campbell rather loftily estimates that it provides an excellent record of English agriculture at the time of its publication, as indeed it does, but it does much more. It describes in great detail the embryonic condition of what was to develop into mechanised farming when the living power of the plough or farm horse was to be superseded, although that was slow enough in developing and took about a century. A horse on a farm, except for leisure riding or hunting the fox, is now scarcely to be found. It is as rare on the farm as it is on the road.

The book was translated into German as *G. H. Andrews: Moderne englische Landwirtschaft . . . mit besonderer Rücksicht auf Deutsches Bedürfniss frei bearbeitet und mit Zusätzen begleitet von M. H. Schilling,* 1855. This seems to be a verbal translation. The illustrations are grouped at the end of the book, an 8vo with a frontispiece showing bringing home the harvest, a not unusual ornament of farming books at that time. Andrews makes some attempt to give the history of the plough but only gets on firm ground with the Romans as might be expected. It would be too much to enter into these details here: it has been done elsewhere.

The Perkins Library has also a *Rudimentary treatise on agricultural engineering,* No. 63 of Weale's Rudimentary Series, a 3 vol. 8vo, London 1852-3. It is an exhaustive attempt to be encyclopaedic and to some degree achieves this purpose. It is laid out in three parts; 1 – Buildings, II – Motive powers and machinery of the steading, the steam engine, portable engines with plates showing the productions of the various makers who flourished or failed at that day. Pt. III dealt with field machines and implements, drainage, the plough, harrows and cutters, rollers and clod crushers and the horse hoe.

Not all of the authors of farming textbooks and similar productions were included in the sacred pages of the D.N.B. But George Glenny achieved that distinction. He was the author *inter alia* of *Farming for the Million*, a copy of which is filed at Southampton. (How I regret some rather acrimonious correspondence with Perkins long before the Second World War!) Of this I only know the second edition, a small illustrated 8vo put out by Eyre & Williams in 1854. It was an ambitious work as might be expected from such a supposedly distinguished person.

Glenny was the editor of a *Royal Magazine* and *St James Archives*, the contributors to which many famous (or then famous!) people like the *Ettrich Shepherd*, Miss Pardoe, Miss Metford, the sisters Strickland. The story of this man could be expanded but this has already been done in the D.N.B. and like good wine needs no trash. There is indeed no complete list of this author's writings, as the British Library Catologue admit. It might almost be asked does it matter?

A brave attempt to produce the history of agriculture (how many there have been, including my own?) was made by one Victor Concalon over the title *Histoire de l'agriculture depuis les temps les plus recules jusqu'à le mort de Charlemagne.* This brave work was published at Limoges, 8vo in 1857. It was apparently one of a series having the general title *Documents inédit sur l'histoire des Gaulois.*

Tables of calculation relating to a particular business, in this context farming, were doubtless of some value to the Victorians: they are today, to state the obvious, superseded by the computer. John Ewart was one who tried to ease the farmers' burden in this way. He prepared *The Agriculturalists Assistant, Principal Rules and Tables*, a small illustrated 8vo published by Blackie & Son in 1857. He was somewhat ponderously pretentious in his preface, arguing that in every kind of business in which manual labour is employed, "by the piece" presents many advantages. Under a properly specified contract, letting work by the piece means constant supervision by the employer to get a fair quantity of work, well done for the cost. The chief purpose of his book was to furnish an easy means of estimating this cost of various operations as well as to ascertain the quality of land operated on. The second part of the book contained rules for computing the value of the piece labour, either manual or by animals, pp. 1 to 78 being occupied by Tables. The latter part of the book discusses manures, their application and weight in relation to cost and in addition most other relevant factors from hay stacks to buildings. He was a very meticulous man and a very talented mathematician.

James Frederick Grant was a surgeon at the Royal Free Hospital who excercised his abilities in producing a book on the *Evil results of overfeeding cattle. A new Inquiry fully illustrated by coloured engravings* in 1858. He protested in his preface that he had long suspected that the English method of feeding cattle was based on a *vicious principle*. His report had he said then lately appeared in the *Observer*, a statement I have made no attempt to check, an impossibly arduous task. Like so many of these writers he treated his readers to a glimpse of the obvious. It must be admitted by everyone that the maintenance of health is a question of higher interest than the removal of disease, though I am rather at a loss to understand this argument. Grant made a succession of complete observations of the obvious in his preface. He remarked that the Smithfield Club had been founded in 1798 for the purpose of finding the best method of rearing cattle used for food. The results of high fattening were marked upon the health of the animal. The beasts became too big for their age and it was likely that fattening deteriorated their internal organs. This is a rather convincing arguement but the man's readers were his judges and today those who read my comments must decide upon the validity or otherwise of his arguement for themselves.

W. Wallace Fyffe, FRSE was the Editor of the *Dorset Chronicle* in 1859 which perhaps gave him some prestige. He was a man with convictions who wrote *Agricultural Science applied in practice ... concise course of scientific and progressive instruction* a duodecimo of 154 pp. published by Groombridge & Son London. He was impressed with a conviction of the value and importance of agricultural instruction as a primary branch of practical education in ordinary schools, where there was a lamentable indifference to the essential facts related to their daily round. No doubt his little book was written for this reason and to supply this lamentable *lacuna*. Fyffe made the self-evident comment that "Man is destined to multiply — not to exhaust but to replenish the earth", a comment that is irrefutable today when the human race has certainly replenished the earth.

Anyone who occupied the post of Professor of Chemistry at the University of Glasgow and was chemist to the Highland & Agricultural Society must presumably (it was not always so) have been qualified to publish a textbook which Thomas Anderson, M.D., FRSE, FCS, did. It was *Elements of Agricultural Chemistry* an 8vo of 299 pp. published by A. & C. Black of Edinburgh in 1860. Anderson had studied under Liebig at Giessen and Berzelius at Stockholm. Anderson carried out many experiments over a quarter of a century most of his results being published in the *Journal of the H. & A. S.*

He was a modest man who declared that his object in publishing his book was to offer a concise outline of the general principles of agricultural chemistry. It had, he wrote, no pretensions to be considered a complete treatise on the subject, being strictly elementary: but he examined the composition of wheat, beans and turnips at different periods of their growth and a number of analyses of soil, manures, plant ashes and oil cakes. His Elements was assessed at the time of its production as not being very original in treatment but nevertheless provided a clear summary of the subject as it was then, dealing with the constituents of plants, soil manuring, farming and manure, sewage manure, vegetable manures, i.e., rape dust and oil cake, seaweed and peat, as well as animal and mineral manures, the rotation of crops and feeding of livestock. He quoted, or perhaps I should say, cited, all the authorities of established reputation whose names are known to all and need no repetition once again.

In all decades, indeed in all times, there are the odd, or possibly numerous, philanthropists whose remedies would cure all the world's evils, if only in their own time and as a result of their preaching. Most human beings are indeed born preachers of one gospel or another. The author of *Our farm of four acres* was no different from any other priest with a gospel to expound. This apostle of the little wrote besides this a larger 8vo volume of 239 pp. no small book, over the title *From May time to hopping*, which Chapman & Hall published in 1860. On this farm Mary's dairy was not a show place but it was cool and scrupulously clean which indeed every dairy must be. This writer also speaks of the "play work" indulged in at hay making, but this production is not a textbook but a romantical exegesis upon country life.

No such criticism can be levelled at Robert Scott Burn who first tempted readers and the makers of fame with his *Year Book of Agricultural Facts for 1859,* a small 8vo of 388 pp. put out by Blackwood of Edinburgh. Burn made the caustic remark in the preface reasonably enough that there were so many Magazines and Journals that nobody connected with agriculture could read them all. The *Year Book* was derived from these publications. It was to be an annual but I do not think it was repeated: I may be wrong. This work opens with a discussion of ploughing and steam ploughing and goes on to the management of various soils, draining, irrigation, manures including gypsum and guano green crop manures, seeds and sowing, the rotation of crops, beet and potatoes, forage plants including sorghum and such novelties as colza. In the second part Burn discussed livestock management and feeding, horses, pigs and the dairy, implements and machines and, rather unusually, legal decisions affecting agriculture. It was certainly a useful and instructive work although the British Library copy remained uncut and consequently unread until I came along.

Burn's next gage to fame and fortune was *Hints for farmers and useful information for agricultural students,* a small 8vo of 171 pp. published by Routledge Warne at London in 1867. This was set up as a popular elementary textbook and contains the usual matter. It would be quite impossible to make a synopsis of the contents of all the similar books put out by enterprising publishers at about the time but I shall make an effort to that end. *Hints* is made up of five chapters the headings of which as is only right and proper indicate their matter, i.e., I.i, Rotations, soils, weeds and seeds, I.ii, The cereal crops and crops cultivated for their seeds, II Root and forage crops, III Manures: farmyard manure is a perfect manure, IV Cattle food, Analysis: statements and calculations referring to stock, V Miscellaneous statements and Tables etc, farming, ploughing, draining, weights and measures.

Burn's next publication was *Outlines of modern farming* in five small 8vo volumes, illustrated, issued by Virtue Bro. London in 1863. It was intended as a rudimentary treatise for students of agriculture. Volume 1 was entitled *Soils, manures and crops*, some 227 pp. Vol II, *Notes historical and practical on farming and farming economy*, 343 pp the introduction to which states that he had given the history elsewhere. Vol III, 211 pp. *Stock cattle sheep and horses*. Vol IV, *The management of the dairy, pigs, poultry.* It deals with the diseases to which the animals were subject. Vol V, *The utility of town sewage, irrigation and the reclamation of waste lands*. It would be a virtue to give a critical note upon the contents of these five volumes and indeed of Scott Burn's other voluminous works, but valuable as this might be for the future historian it is not possible here, if only for reasons of space.

This was followed by *Notes on an agricultural tour in Belgium, Holland and the Rhine,* an illustrated 8vo of 241 pp. published by Longmans in 1862. It described the pleasures of the tour and the strangeness of the people so different in air and clothing to the familiar English. On farming the cropping in various places is set out, the treatment of manure, the general principles of Flemish husbandry, the implements of Flemish design, pictures of and discussion of the Flemish ploughs, the cultivation of flax and colza, the condition of the peasantry, and the College of Bonn Poppelsdorf, a sufficiently comprehensive production.

After an interval of nearly twenty years Scott Burn published *Outlines of farm management and the organisation of farm labour, treating of the general work of the farm; field and livestock; details of contract work; specialities of labour; economical management of the farmhouse and cottage; and their domestic animals etc., etc.,* with numerous illustrations, an 8vo of 272 pp. published by Crosby Lockwood in 1880. This even deals with female education and contains paternal moralisings about the cottager's wife and the man himself not to speak of the cottager's pig and poultry, rabbits and pigeons, an exhaustive and very moralising work.

Finally for Scott Burn there was *The practical directory for the improvement of landed property, Rural and Suburban and the economical cultivation of its farms,* a large work of 603 pp. with 77 plates bound at the end of the volume, depicting housing, machinery, draining and so on, in addition to various woodcuts in the text. It was published in 1881.

A very much lesser light was Thomas Bate who wrote a 28 pp. 8vo pamphlet issued by G. E. Thomas of Chester entitled *The food of plants. A plain treatise on manures,* the only copy being known to me being filed at Southampton and therefore not immediately accessible to me. There is no copy in the British Library, so this doubtless impressive work must remain to fame in its mention at this place.

Mrs Fergusson Blair of Buthayock was, so she herself says, a successful breeder of poultry and the winner of more than 300 prizes, no mean achievement indeed! Her book *The Hen Wife her own experience in the hen yard*, or so she says, was an 8vo of 192 pp., adorned (and the word is really applicable) with many coloured illustrations of different breeds. The first edition of this immortal work was published by Thomas C. Jack of Edinburgh. This treatise went into no less than eight editions, which fills me with awe and admiration. Mrs Blair said that it was written because she had been asked to write it, and she felt she had no right to withhold the knowledge she had. It had a firm foundation on the winning of prizes, as already mentioned, and also her conscience told her that she had no right to withhold the knowledge she had: but this is a minor subject though very important for the production of eggs and the edible carcasses of the hens.

All bibliographers are confronted with the need to make decisions (what man is not?). Every now and then one must make up one's mind whether or not to include a certain book or not. In 1861 Slatter & Rose of Oxford and Week Bell & Dalby of London issued a 44 pp. illustrated 8vo *The agriculture of Berkshire*. A subsoil map of the county formed the frontispiece. A rather obvious remark is that the Thames was not the principal carrier of goods before the Great Western Railway was constructed. Another is that the riparian meadows were good if not flooded too long, the subject of floods being carefully discussed. Clutterbuck followed Manor (Gen View sci. Vol III) in his division of the county into Vale, Hill and Forest, as nature prescribed. Clutterbuck refers to Caird's description of Sir John Conroy's farm and reclamation work, his modern farm buildings and appliances, Shaw Farm Chiefly Park. It was grazed by 200 head of cattle, mainly shorthorn cows with some Channel Island. A detailed tale of the Royal

Farms at Windsor is provided.

In general the breeding and feeding of sheep was, he said, the ruling feature of Berkshire farming, especially between the Vale and River Kennet. The Vale of White Horse was of exuberant fertility, soils being carefully described. A few farms and their operation gives a good general idea of the state of Berkshire farming at the time of writing.

Clutterbuck also put out *Agricultural Notes on Hertfordshire*, a reprint of an essay that had been published in the Journal of the R.A.S.E. of 1864. It depended to a large extent on Young's and Mavor's writings, which is all that need be said.

To my mind it is an odd caper to seek fame under the guise of anonymity, or of what is a rather curious cast of mind which looks like Mr Facing Both Ways, the wish for recognition and the desire to conceal one's literary activities. Such an odd ball called himself Peter Dryland. Maybe he wanted to hide himself because he was uncertain of his qualification, or possibly because he did not want to claim authority by reason of his social and professional status, he then described a rather improbable name on the last page of his book: certainly an intriguing person. Why these unnecessary divulgations? Who can say a century later? But curious!

The opening sentence is pompous and rather amusing. "There is no work on Agriculture extant as yet that gives a glimpse of the future to a gentleman who intends to farm either as a young man renting land or as an owner of property who means to farm his own farm." This is pompous and of course quite erroneous. Dryland, if he was properly named Dryland, admitted that his readers might not have the patience to finish reading his book, only an addition to the numerous works which are only read and understood by the initiated. Farming is as easy as eating and drinking. Have we not seen this idea displayed in "Farms of 4 acres" as well as on farms of 450? He then discusses the Royal Agricultural College at Cirencester, and some criticism of farmers who take pupils. If you dream of anything, he wrote, dream of the steam plough. Thereafter he goes on to discuss landlords, tenants, bailiffs, cotters, agents, and the processes involved in improving property with a list of essential stock and implements. This rather pompous production could be largely discussed but that might be egregious. On p. 292 Dryland wrote "Now to conclude these trite remarks" (unbelievable false modesty) he trusts that the reader had as much pleasure in the perusal of this light volume "as I Peter

Dryland have experienced in occupying some of my leisure in penning it."

The title of this rather intriguing book is *Farm life or sketches for the country* a London 8vo of 292 pp. put out by Saunders Ottley London in 1861.

Thomas C. Fletcher contributed to one of Routledge's series of Books for the Country. Its title was *Scientific farming made easy; or the Science of Agriculture reduced to practice*. I have been unable to trace the issue of the first edition but the second, revised, corrected and enlarged came out in 1861, being also put out in New York. The R.A.S.E. Library has a third edition, also "revised, corrected and enlarged" an 8vo of 1864. The introduction is a piece of, I fear, prideful self denigration. "Another farming book." What can authors know of farming! Don't tell me of a man delivering his precepts on Agriculture from his study surrounded by books! Without experience, propounding new facts, theories, and laying down new law when in fact they are in most part mistakes copied from other wiseacres preceding him. This author gives his object as being to list existing theories by practice and give to the Agricultural Community the benefit of those incurring the cost, and trouble of investigating the discoveries of others, and the office of a lamp to enlighten its path. Science, Fletcher wrote, has pointed out simple methods by which a farmer can increase his produce and augment his profits and he then proceeds to describe what the farmer should do in producing crops and animals and their products. There was, he said, some resemblance between animals and plants. Animals possess powers requisite to enable them to support their state of being, varying according to their nature and species. Precisely the case with plants: they are amply endowed with the powers needed for preserving vegetable life. Having laid down these principles, Fletcher then goes on to detail the methods that ought to be used. Little more need perhaps be said except that the book more or less fulfils the plan the author had set out for himself. Few writers can do more!

John Gamgee was a very different type of man. He was Professor in the then new Veterinary College at Edinburgh. He wrote two books, the second ten years after the first, which was *Dairy Stock; its selection diseaes and produce. With a description of the Brittany Breed*. This was an illustrated 8vo of 316 pp. having an interesting pictorial frontispiece depicting a Brittany cow. It was published by T. C. Jack of Edinburgh and Hamilton Adams in

London in 1861.

In the preface Gangie protested that he had stated revised opinions and advanced his own with all freedom and he trusted with fairness. The breeds of excellent milk cows were, he said, very numerous and to be found in every county. It is remarkable, he said, how much prejudice retards improvement in all things, a statement which was as precise and true then as it is today. Breeds of excellent milk cows are very numerous and to be found in every county. Gangie depended very largely upon Guenon.

There were, he wrote, three heads: (1) The Country dairy cow, (ii) the Town dairy cow and (iii) the Family dairy cow. In many counties dairies were stocked to some extent, occasionally entirely, with home bred beasts of greater purity than in the dairies for the supply of large towns, e.g., the Ayrshire, Galloway etc. Part II of the book deals with Brittany and its purple illustrations of a dairy then which is very nice, and one of a Brittany bull. Gangie was a protagonist of the breed of this then foreign country. The admixture of foreign blood into our dairies has provided incalculable value, or so he said, but it would be easy to expand this topic, a temptation I must resist.

Although the condition of the agricultural workers, the helots of a temporarily flourishing occupation, was deplorable, their housing as deficient as their wages, there was no lack of "genteel" patrons of the country life, a self sufficient one if ever any life is self sufficient. Harriet Martineau was a prominent and respected member of the class ridden social life of her time. She had been or should I say was a contributor to Dickens' *Household Ways* but what concerns us (me) here is her product *Health Husbandary and Handicrafts* a sonorous title to a rather philanthropic work though whatever results it may have had are yet to be discovered. It was an illustrated 8vo pf no less than 583 pp. published by Bradbury & Evans, London 1861. It would be easy to expand upon her ideas which indeed to some degree must be done. Who would ignore Harriet Martineau? She was pedestrian and pompous, proclaiming that "It can give me nothing but pleasure to join in the endeavour to make useful the results of a long experience (she does not say 40 years as so many others did) and observation on the homely realities of life." This is criticism of society and includes a chapter on the rural labourer. Her exegesis on his health and so on which may be called the "hou ha" of that time, the praise of rural life and the neglect of the man who did the work. Chapter viii discussed the rural labourer but was really a general exegesis on the classes Soldier and Sailor and the Aged.

This lady also produced *Two letters on cowkeeping addressed to the Governors of the Guilteros Union Workhouse,* an 8vo of 20 pp. which need not be discussed here. Her ultimate aim seems to have been not procuring profit but comfort by providing her own subsistence. "If without loss," she wrote, "I can provide myself with hams and bacon fowls and eggs, vegetables (except winter potatoes) but I am told I cannot grow enough to last the year." An apothegm is the more manure, the more green crops, the more stock, which is a literary piece of rhodomontade but is nevertheless rather strikingly apposite.

A little book, or should I say pamphlet, by W. F. M. Pocock is filed or was in 1864, in the Bath and West of England Society Library. It was *Farm life or sketches for the country.* It was a 12vo of 31 pp. I have not seen it for this is its only provenance. There is no copy in the British Library so far as I could discover.

Alexander Smith of Falkirk was quite a different sort of man. He wrote *Agriculture; a poem in 16 books* some 281 pp. 8vo, published by T. C. Jack of Edinburgh. He was the son of a lace designer but proved an indifferent successor to his father in this trade. In 1854 he obtained the post of Secretary to Edinburgh University, a position which gave him leisure enough to allow him to write for newspapers, magazines and encyclopaedias. His poem has never been esteemed as more than jejune. His preface is rather amusing. "Things that are known I sing . . . I pretend discoveries none to make . . . yet urge I may neither does gainsay. Hence all with safety may then pages read". His poetic diatribes discuss earths, soils, manures, crops, culture and tillage. The thing forms good and interesting reading and is time consuming for that reason. I cannot assess its agricultural value or imagine what sort of impact it made upon what was doubtless a largely indifferent public.

A farmer who liked indifferent cattle for their cheapness would be a foolish and rather unusual man. R. D. Pringle did not fall into that catergory. He was the author of *Cattle Management, being directions for the rearing of calves* . . . an 8vo of 64 pp. dated 1862. Pringle, to whom we may return later, did not like cheap beasts. Many thousands of inferior cattle are bred, though it is quite as easy to breed the better or best sorts. This was true, or so he said, of all breeds. This carelessness was a waste of time and money he declared. The book went into a 3rd edition edited and revised by James Macdonald, 8vo Edinburgh 1886 to which I shall return in the appropriate sequence.

Another Scot was John Wilson of Eddington Mains, Berwickshire. He was much more expansive than Pringle in his book *British Farming; a description of the mixed husbandry of Great Britain* an illustrated 8vo of no less than 569 pp. issued by A. & C. Black of Edinburgh in 1862. The plates are bound at the end of the text in a group and show plans of farm layout and of animals, horses, sheep, pigs and so on. The preface states that the author having contributed to the (then) new edition of the *Encyclopaedia Britannica* the articles, Agriculture, Dairying and Draining, he has now at the request of the publishers embodied these in the following treatise with considerable corrections and additions. As an introduction Wilson wrote an Historical Summary of ancient, medieval and foreign agriculture in which he discussed Early British Agriculture to the end of the 15th century. Little need be said here about Wilson's description of contemporary farming methods although his precepts were clear, but not exactly concise. He must have read the books for he gives a rather precise and pleasant bibliography of agricultural writings from that of Fitzherbert to the only approximately farming books written by Ray and Evelyn. I do not think it essential to the present purpose to enter into minute detail about the contents of this book which are possibly sufficiently self evident: they are in any event, what may be called normal to the genius.

A pamphlet filed in the Highland & Agricutural Society's Library was written by one James Drummond. Its title is *The true cause of the Vine Disease and the Potato Disease* Edinburgh 1863. As this seems to be the only recorded copy of this profound work I have not been able to read it and cannot therefore criticise or otherwise comment upon it.

The Secrets of Farming were disclosed by one John Large in his book of that title, an 8vo of 155 pp. published by Allen of London and Bull of Swindon in 1863. It was a subscription publication approved by the Duke of Beaufort which undoubtedly gave it a cachet in those days of a very different social outlook from that of the present day. In his preface Large apologises for any abruptness that may happen to appear in his writing, and humbly submits it to the public as indeed do all aspirants to literary fame. He claims to have had 16 years experience on almost every description of soil —

a statement that was possibly an indication of a very frequent change of venue. The book is set out in the form of an alphabetical series under subjects as Large selected and nominated them. One of his major subjects was the most economical method of feeding and keeping farm horses, a subject on which he owned a certain expertise if I am not mistaken, but perhaps this is not the place to examine his ideas.

Since an exhaustive life in the form of *Memoirs and Letters* was edited by his son Arthur H. D. Acland in 1902, a work of 421 pp. adorned with a frontispiece and plates, it is unnecessary here to do more than mention that the 11th baronet (1781-1871) was the author of a good many brief pamphlets on a variety of agricultral questions, before which he had written in collaboration with William Sturge a full scale study of *The Farming of Somersetshire* an 8vo of 179 pp. put out by Murray of London in 1851. A loose insert of two maps, on one folded page, showed the physical and geological condition of the county. Sturge's essay is separate. It was done to compete for the prize of £9.00 offered by the R. A. S. E. This is a careful survey and of great value from an historical point of view. It described Downside College which had a farm of 308 acres. Acland made an interesting remark to the effect that when we call to mind the services rendered to agriculture by the Monks of the Middle Ages even in Somersetshire it may incite less suprise that this is one of the best appointed and best managed farms in the county, a statement that is followed by a description of this superlative undertaking. Much land in the county, he said, then needed reclamation by which he only meant good farming.

Sturge's *Report on the farming of Somersetshire* was written to compete for the premium of £50.00 offered by the R. A. S. E. It was unsuccessful but appeared soon after Mr Acland's whose Prize Report was published in the Journal R. A. S. E., vol xi, 1850, pp. 666-763, illustrated by two maps. Consequently the book is but a reprint of this essay, which was illustrated to an informative extent.

His first aim at fame, though possibly his intention was only to stir up public opinion, was *A Letter to W. Miles, Esq., M.P., containing a proposal for the establishment of annual agricultural writings to behold successively in different towns in the West of England with Mr Miles' reply,* a 16 pp. pamphlet issued by F. May of Taunton in 1850.

Seven years later Acland wrote a 60 pp. pamphlet with the title *The education of the farmer viewed in connection with that of the middle classes in general.* It was issued by Ridgeway in 1857. In the same year he put out *Meat, milk and wheat, an elementary introduction to the chemistry of farming to which is added a review of the questions at issue between Mr Lawes and Baron Liebig,* reprinted from the Bath and West Agricultural Journal, a 97 pp. 8vo, put out by Ridgway in 1857. This was in three parts, I. Introducing principles . . . agricultural chemistry not intended to teach farming but to explain it. II. The application of principles to farming, but readers need not expect too much because of our imperfect powers of analysis and of life. He professed to explain the effects of dung and guano and described or rather referred to the experiments of Lawes and Gilbert. III. He discussed Liebig's controvesy with Lawes.

His next publication was it seems an 8vo pamphlet of 52 pp. on *Agricultural Education, what it is and how to improve it, considered in two letters to Sir Edward Kerrison Bart. With an appendix containing the results of enquiries amongst farmers and schoolmasters,* printed by W. Clowes, London 1864. This need not be discussed here as I have written a full scale essay on *Agricultural Education* elsewhere.

Acland's next (?) publication was *An introduction to the chemistry of farming specially prepared for practical farmers with records of full experiments,* the 2nd edition of which was put out by Simpkin Marshall in 1892, the first edition having appeared, so far as I know, in the previous year. He made the caustic comment that it was needful to translate chemical language of the present day into the mother tongue. He was not very modest. "I will then hope that I shall have given some tangible meaning to some technical words", a hope to which I subscribe.

Acland's works are a puzzle to me but I hope that I may have seen them all and noticed them casually and caustically here. The last seems to have been *Review of agricultural experiments containing criticisms and suggestions. With a letter to the Earl of Chichester.* The first part reprinted with additions from the Bath and West of England Society, 3rd ser 16, 1885.

The first edition of William Bland's *Principles of Agriculture* was put out from Hartlip, Sittingbourne, Kent on 28th February, 1827, the second (?) a small 8vo of 127 pp. was published by Longmans Green in 1864, an interval of some 37 years. Modestly enough the preface to the second edition states that he had been induced to issue it "Knowing that it is not uncalled for", and because he felt it might be of use. His Honorary Secretaryship of the Sittingbourne and Milton Agricultural Club gave him some kind of authority. Since his first edition an immense advance had been made by the aid of science in the accessories to agriculture, particularly in machinery and in the introduction of new manures, a good enough reason, or so he thought, to publish a new or second edition.

The book is well planned, as I said of the first edition, proceeding through the subject from the analysis of plants, the sources of the food of plants, what causes fertility, fallow etc., etc., throughout all the processes of soil preparation, its enrichment by manures and tillage, the alternate husbandary, crop rotations and so on. It concludes with Hints to young agriculturists. He advised that agriculture should be carried on with a view to profit, which seems a rather unnecessary exordium, being something every farmer is well aware of. It was of course excellent advice, if perhaps unnecessary.

A writer who was the Principal of the Royal Agricultural College at Cirencester and an M. A. of Trinity, Cambridge, might be expected to be readily an exponent of *Agricultural Education* delimited in six lectures by himself and others published by Longman, London & E. Bailey, Cirencester, in 1864, an 8vo of 167 pp. The lectures were all by then distinguished men, the first of course by the Rev. John Constable himself, Chalmers Morton, A. H. Church, Professor of Chemistry at Lincoln Hall, Oxford, John Bayldon, Professor of Botany, A. J. Murray, Professor of Veterinary Surgery. This was followed by a description of the College and the management of its farm in both arable and animal husbandry. A further co-operative work in which Constable played a leading if not initial part was *Practice with Science; a series of agricultural papers* in 2 vols., 8vo, published by Longmans 1867-1869. Its contents can be assumed by the informed and intelligent reader, and not be gone into with beyond that opinion.

The disposal of waste products was always a problem in the growing cities like London and those others of the industrial age, and a good many pundits wrote up their opinions of how these waste products should be disposed of or used to fertilise future food. One of them was *A letter to John Thwaites, Esq.*, a brief 23 pp. 8vo put out by Edward Stanford in 1864. It did not in fact provide any novel ideas: it was quite the usual thing about town waste, which recurred at intervals as new protagonists of the sublime purpose of using town wastes as manure rose up and voiced their opinions.

Duncan George Forbes MacDonald, 1823 (?) to 1884 had at least a resounding nomenclature: and he wrote several books. He was a compulsive writer who did not confine himself to farming and its topics but is described in the D.N.B. as an agricultural engineer and miscellaneous writer who "early devoted himself to the study of farming on his father's extensive glebe" but his other extensive and widespread activities (in a geographical sense!) need not concern us here here, where our concern is with his writings on farming and its problems. The first of the gages he cast at fame in this subject was *What the farmers do with the land; or Practical Hints for their and its improvement,* the 2nd edition of which (I have not been able to trace the first) was a 51 pp. 8vo pamphlet put out by H. Adams of London in 1852. This was evidently well received for it was published again by the author as *Practical hints on farming and estate management*, the second edition of which was a 73 pp. 8vo printed by Edward West in 1864. MacDonald made the profound observation that agriculture is essential to life but the farmers were slow to leave the beaten track. The Royal Agricultural Society has worked invaluable good. The local associations have also in their way been useful for the promotion of agriculture (a somewhat patronising remark). Books on the subject were increasing in number (as I very well realise!). One Alison has said the husbandry of Flanders in our plains, that of Tuscany on our hillsides would easily raise food enough for double the present population. He cites various writers, Smith of Dunston, Professor Johnston, *Lectures on Agricultural Chemistry,* Mechi. He gives figures of yields in England and Scotland which were only estimates, of course.

One of his criticisms of practice was that too many horses were used in the plough, six each attended by a man or a boy besides the ploughman. He recommended the two horse plough as being just as good: too many horses made the furrow awkward and irregular; too light ploughs made only a scratching of the surface. But on the whole this could be described as more economic and argumentative than technical.

There is one remark that ought to appeal to modern farmers. "The shortness of the ridges is a defect in the economical view of the English system which I cannot pass over without recommending an improvement. In many cases (he means on many farms) no doubt an advance in this respect would necessitate a change in field boundaries and the uprooting of vast quantities of hedges as it is clearly impossible to make long ridges in short fields (as may be seen today)." The mania — it may be called — which prevails in many counties for small fields encompassed by hedges is absolutely suicidal. Many counties like Devon have fields that are too small. Modern arable farmers would no doubt agree. Away with hedges and hedgerow trees. Make a prairie that can be "done" with modern machinery. How novel are modern ideas and practice? MacDonald gave some interesting figures in support of his argument, and further discusses cattle, horses, drainage and the improvement of waste, labourers' cottages etc., and advertised his other books in the end papers.

In 1872 he put out his ideas on *Cattle, sheep and deer; containing also remarks on the game laws and grouse moors,* the 3rd edition of which (I have not seen the 1st and 2nd eds.) has a frontispiece as issued by Steel and Jones of London in 1872. The R.A.S.E. library has or had a 10th edition of *Hints on farming,* said to be a London 8vo of 1868. I have not seen this! This writer also wrote on *The grouse disease, its causes and remedies,* an 8vo of 1883, but not precisely a farming subject.

He was by no estimates a modest man. "I have", he wrote, "prosecuted my calling as a civil engineer, agricultural engineer and estate agent with success. Under my immediate direction and superintendance upwards of a million sterling (a vast sum in those days) has been expended on the improvement of agricultural lands alone." Need more be said on his analogous achievement and literary production.

Henry Woods of Merton, agent to the Rt. Hon. Lord Walsingham, gave a lecture which was afterwards published as a pamphlet, *The breeding and management of sheep. A lecture delivered to the Wayland (Norfolk) Agricultural Association,* an 8vo of 47 pp. put out by Day, Son & Hewitt of London in 1864. The British Library copy was uncut when I examined it. The Perkins Library has apparently another copy, *A lecture on the breeding and management of sheep delivered before the Wayland Agricultural Association,* a second edition, London 1864. This was translated into German as *Über die Züchtung . . . des Fleisch-schafe. Ein Vortrag . . . aus dem Englischen übertragen*, 1865, an 8vo. Woods was to pursue the subject with *A lecture on the diseases of sheep with plain and practical directions for their treatment*. Revised and enlarged, an 8vo of 1873. But informed and informing as he was on this subject it was not his major claim to fame or possibly should I say inclusion here. That was *Ensilage: its origin, history and practice with an account of experimental trials and results at Lord Walsingham's Home Farm*, an illustrated 8vo of 63 pp. published by Stevenson & Co of Norwich and Ridgeway of London in 1883. In the following year, 1884, Woods gave a lecture to the Institute of Agriculture (an Institute I know no more of than its mention here), *A lecture on ensilage; its influence on British Agriculture* enlarged and revised with appendices etc. The practice of ensilage was, said Woods, very ancient — the practice, that is, of pressing produce in structures from which air and moisture are excluded. Thorold Rogers says it was known in the 5th century before the Christian era. It was mentioned by Euripides and described by nearly all the Latin writers on agriculture (he gives no bibliography!). More, he furnishes abundant confirmatory evidence in quotations from the old historians and scriptural records. . . not the least doubt that among the ancient Hebrews the practice of burying or covering over their corn was widespread. . . provision against years of famine. Other evidence can be gleaned from Egyptian paintings. . . grain stored in vaulted chambers. The antiquity of silos and of a certain form of ensilage is established. It had long been practised in the continental countries, Germany and Hungary in particular. His arguments are supported by notes of milk yield and butterfat content on Lord Walsingham's estate. Woods also wrote a more particularly veterinary work, *Lecture on abortion and mortality amongst ewes*, obviously of interest to sheep farmers, but rather more properly to be placed as a veterinary rather than farming treatise.

One John Fisher, of Carhead Farm nr. Crosskills, was undoubtedly a generous man. He protested that he did not want to keep his knowledge to himself and so he wrote a 14 pp. pamphlet on *The breeding and management of pigs*, no very exhaustive treatise. It was an 8vo published at Skipton in 1865. "Nothing but a deep sense of its (pig breeding) great importance, and the hope I may be able to contribute something towards its advancement" had made him publish this didactic production. It was indeed a paper he had read to the Newcastle on Tyne Farmers' Club in April, 1865. Another stimulus to writing it was the importance of the subject as a source of profit, a circumstance that must have appealed to his readers. Fisher ends on a homely maxim which protested "It is a poor house where they never kill a pig" but this trifle is in such small print that it is difficult to read and the information is therefore difficult to pick up. About 1870 (?) Fisher put out a 2 page leaflet *Book of points of the pig* setting out points for each part of the animal. I suppose this must have been useful to somebody.

I think a mere mention of a pamphlet of 21 pp., an 8vo adorned with a frontispiece and plates will suffice for *Cattle food adulteration; being a guide to the farmer in the purchase of oil cake,* 21 pp., 8vo, written by William Henry Harris. I could not presume to comment upon it!

Botanical works are not strictly farming books, which are what my title claims as my subject but they have clearly a contingent interest for the readers of farming books, which is my excuse for including such works by Thomas Croxon Archer, of the Edinburgh Museum of Science and Art. Strictly speaking, his early work if mentioned at all should be in the previous volume of this bibliography. It was in 160pp., published by Reeve & Co, London in 1853, *Popular economic botany or a description of the botanical and commercial characters of the principal articles of vegetable origin used for food, clothing, tanning, dying, building,* etc. The preface was signed as from Higher Transmore Cheshire May, 1853. He subscribed to what is now an established belief, or so it seems, that cereals originally grew wild somewhere in Russia or Asia. Though not precisely an agricultural book it is interesting in its mildly outmoded theories of the origin of this or that. It even mentions spices, nuts and so on. An abridgement of this work was published in 1854 by Lovell Reece, London, a 16mo with the title *First steps in economic botany.* This was followed in 1857 by an 8vo *On the study of botany.* It was one of a series of inaugural lectures delivered at the Liverpool Ladies College during its first session in 1856 and was published in 1857. It is educational, as is proper for such a thing intended for student ladies but not farmers. His next publication was *Profitable plants, a description of the principal articles of vegetable origin used for food, clothing, tanning, dying, building medium, perfumery* etc., a small 8vo of 359 pp. with a frontispiece and 20 pp. of coloured plates issued by

Routledge, Warner & Routledge in 1865. It can hardly be described as an agricultural book. Archer also wrote on *Wool and its applications* in 1876 but this is not a farming book, whatever generous bounds may be set to the subject.

John Bailey Denton was not a farmer but a land agent in Gray's Inn Square, where he executed his business as is or was no doubt appropriate to anyone living in that place for example Samuel Butler, who wrote *Erwhon* amongst other fascinating *Selected Essays* printed in 1927, a work I have found interesting for at least half a century; but to return, Denton's major work was *The farm homesteads of England. A collection of plans of English homesteads . . . to illustrate the accommodation required under various modes of husbandry,* the 2nd edition of which was an illustrated folio of 188 pp. which Chapman & Hall published in 1865. I am not qualified to criticise Bailey Denton's plans.

Bailey Denton had previously put out two other things. One was an *Outline of a method of model mapping* which I have not seen and the title of which does not disclose its contents – at least to me. It was a London 8vo of 1841. In the following year Bailey Denton had put the question *What can now be done for British Agriculture? Answered in a letter to Philip Pusey, Esq., M.P. advocating a general system of drainage with a profitable distribution of the surface and drainage waters and the refuse of towns*, an 8vo, London 1842. By the improvement of the land, he wrote, is to be understood a greater salubrity of climate, more congenial to the vegetation as well as the animal kingdom and an increased and more valuable produce both of tillage land and pasture, which was a simple profit and loss argument. He quoted the census which showed that 7/9ths of the population depended on farming: only 2/9ths on manufacture. An argument that does not impress me, perhaps because it is quite unsympathetic to my cast of mind.

The problem of sewage disposal and its possible use was one that occupied a great deal of thought or attention as the condition of houses and cellars became more olfactory with waste products human, animal and vegetable. So John Bailey Denton, whose business as a land agent must have brought him into close contact with the solids and smells, exercised himself in producing in addition to his other output, a 66 pp. 8vo pamphlet on *The sewage question. Letters with appendix explaining shortly the several processes adopted for the treatment and utilisation of sewage with special reference to the solution and preparation of land for irrigation and for intermittent downward filtration*, an informative but rather incomprehensible title, issued by E. and F. N. Spain, London 1871.

Bailey Denton explains the necessity for under draining all irrigated lands and so on and so on. This sewage disposal and utility question was widely discussed at the time, a great deal of print being used in the process possibly without very much result, but useful as the product was, as the wharfingers who took it to the land well knew, it seems to have been rather freely discussed, although it could not have been over important at that time.

Samuel Copland adopted the *nom de plume* of the Old Norfolk Farmer (of the Mark Lane Express) in writing an 172 pp. illustrated 8vo work for the County Library and Family Circle Books, *Wheat its history, characteristics, chemical composition and nutritive properties* put out by Warne, London 1865. Its title is an effective guide to the contents of this work, a discussion of which would be rather unduly repetitive and say no more than would the syllabus of all the other similar works.

In the following year 1866 Copland organised and made substantial contributions to *Agriculture, ancient and modern, an historical account of its principles and practice exemplified in their rise, progress and development*, 2 vols., 4to, published by Virtue, London.

One unrelated consequence a century after their publication is that I have to say something about the books here, part of my self-appointed employment. Perhaps I may be permitted to say that I have for long had a major admiration not only for Copland but for his collaborators like Hoskyns, the last of whom has always enjoyed my admiration. He has been so long dead that he cannot appreciate this honourable distinction. Alas! However I must proceed!

The book was copiously illustrated with excellent engravings of crops, implements and stock. It was characterised by a genial but engaging conceit and was indeed a remarkable work although as every encyclopaedia must be, was a collaborative undertaking with a remarkable flavour of the history of farming, growing crops and herding cattle from the very earliest times as the subject was then understood.

One remark, not about farming, intrigues me. It was that recent discoveries have given rise to the opinion that the creation of man was not coeval with the earth, but he nevertheless accepts the biblical story and asserts that man was only 6000 years of age and therefore (though he does not follow it) farming could not be any more ancient, Cain and Abel being the farmer and the herdsmen as the Hebrew Scriptures put it (with the authority of their imagination!). This encyclopaedia makes the assumption that Agriculture is the embodiment of all the physical sciences, which might be thought to be a rather overwhelming statement if it could be accepted.

Our Victorian ancestors certainly were learned in the terms of their day and were interested in origins. For example Copland discussed *Wheat, its origin and history*, an inexhaustible subject even today. He subscribes to much Bible history and to the classics, referring to Rees' *Encyclopaedia*. Cultivating, he wrote, was extended by the monks who purchased large tracks of land then lying waste: but such a work could call for a whole treatise of criticism or comendation for which there is no place here. It may perhaps be said that Copland, or his writers, discussed some subjects that are today quite exhausted. The technical instructions in the book follow the seasons with appropriate guidance for the reader. Pests and diseases are described both for stock and crops: housing and marketing are not omitted.

The Salt Chamber of Commerce of Northwich does not seem a likely organisation to be in any way concerned with farming, but all the same it was, and it offered a prize to literary aspirants who cared to write about this mineral. The award was made to Robert Falk for his essay, *The use of salt in agriculture, prize essay published by the Salt Chamber of Commerce of Northwich. First prize awarded to Robert Falk, second prize to . . . T. H. Phipson*. It was an 8vo of 56 pp. published by G. J. Poole, Liverpool and Simpkin and Marshall of London in 1865. I am indebted to the Perkins Library for this reference for the British Library copy seems to have been destroyed by bombing during the Second World War. Consequently I cannot discuss Falk's theories.

Farmers are not the most trusting of people so an author who was asked by the local pundits to write up his ideas so that they could take his advice must have been duly flattered, supposing this pretence was not just an advertising gambit. Such a possibly devious man was Alfred George Lock, the first edition of his phamplet *Agriculturists their non superphosphate makers* being published by T. G. Gutch of Southampton. It was a 26 pp. 8vo supplied with a frontispiece and other plates, a second edition coming out ten years later in 1872. Lock criticised the adulteration of guano, no doubt with some justice, and based his assertions on the accepted authorities. He praised the use of bones in the impoverished soils of Chesire and other cheese exporting counties, their use being a minimum boon and duly appreciated by the possibly satisfied dairy farmer. With increased solubility there was evidently an increased production power of the crop, by which he no doubt meant milk and cheese output. This modest man completed his lecture by appending reprints of testimonials he had received.

A copy of his 44 pp. 8vo on *The true art of manuring: a reply to Mr Spooner's lectures at Winfrith and Botley* is filed in the Perkins Library but not in the British Library with the consequence that I have not seen it. Harrison of London printed it.

In 1865 the disposal of London's waste products was a formidable problem, not easily solved, but very easy to write about. One of these lucubrations was printed in 1865. It was *The agricultural value of the sewage of London examined in reference to the principal schemes submitted to the Metropolitan Board of Works. With extracts from the evidence of chemists, engineers and agriculturalists,* an 8vo with a frontispiece and 78 pp., issued by E. Stanford of London. The R.A.S.E. Library has a much earlier *London Sewage Co. Report upon the plans proposed for rendering available the manures obtained in the sewage water of the metropolis,* 1845, but this is not in the British Library. The subject was recurrent and I doubt if I have been able to pinpoint all of the related pamphlets and so on.

I have to include here, but cannot discuss Magne, who produced a work *The Home and Foreign Agricultural Miscellany. The cow and the dairy. Pt.I, The cow*, provided with a nice frontispiece and said to be adapted to English practice, which I regard as a doubtful qualification for this production. It was profusely illustrated and supplied with sketches and tables, but was decidedly of a difficult approach to its subject, at least for English readers, one of whom is myself. Whether it had any influence or effect on English practice is to seek. I certainly cannot define it – if any!

There is certainly a debt to Walter Frank Perkins that assiduous collator of bits and pieces, as well as major products, of farming advice, if it cannot always be said to be knowledge. Fugitive as so much of this work, pamphlets and so on, was, it was certainly worth collecting and when it was being collated was by no means costly.

One of these productions which is not in the British Library is by one Theophilus Redwood. It is of course no more than a printed lecture entitled *A lecture on chemical aids to agriculture delivered to the Cowbridge Farmers Club,* a 31 pp. 8vo, printed by J. E. Taylor of London in 1865. I have not seen this!

The task of a bibliographer is difficult, tedious and possibly unnecessary, though I do not think so or I would not have undertaken it. Someone who signed with the initials IEEB wrote a 171 pp. 8vo published by Horace Cox in 1866 which seems to have been the third volume of a series of publications arranged under the general heading of the Field Library. The title of this production was *The farm garden, stable and aviary* and its claim to notice was made by the assertion that it was valuable to Country Gentlemen, Farmers, Gardeners and their professional brethen. Although farmers appear in the title page, the book was really intended for the country gentleman who wished to be instructed in the best way of living his life. Such arts as horse shoeing, riding, veterinary information and so on are carefully set out. What would not have, I suppose, appealed to the working farmer was such curious information about varieties such as growing walnuts and the cure of decorative poultry. A curious but interesting work, informative indirectly about the way of life of the wealthy living in rural areas.

A Hampshire Farmer, one James Neville Fitt, was of a very different style. He wrote, or at least so it is said by Perkins, *Hearths and Homesteads; a series of papers on subjects connected with agriculture,* a sufficiently encylopaedic title for a work of 93 pp., 8vo, published by J. Humphreys, London 1867, but the attribution is only a mild suspicion. Fitt, somewhat doubtfully, is attributed author to a small production *Hunting, Steeplechasing and Racing Scenes,* a square 8vo of only 34 pp. but having the distinction of having illustrations by Ben Herring. The book is included only on the excuse that hunting was a rural sport, if I may put it on such a low level, when it was for part of the year the daily occupation of certain landed gentry.

The odd details that came to light are largely what adds interest to the somewhat tedious tasks of bibliography! William H. Heywood adds something to this. He wrote *High farming. How far is it expedient? Let nature answer,* but he presumed to offer some kind of answer to his own question in his pamphlet. In this case nature was Heywood in a 42 pp. small 8vo issued by G. Phillipson and Golder of Chester: Griffith and Farmer; W. Ridgeway, London in

1867. This pamphlet had a cachet, being dedicated by permission (mark you!) to the Rt. Hon. Lord Egerton of Tatton Chester. It was distributed by the usual publishers of such things.

This author professed the highest principles. He was activated by a profound admiration of his Lordship's course in the Management of Large Estates and in dealing with a numerous tenantry. Although few of our Victorian ancestors were modest men, much the opposite, Heywood professed some self deprecation in his preface which runs "I feel that a short comment is necessary to explain the publication of such tenets at a time so adverse to their being generally adopted as for instance the advocacy of an extension of grazing and the diminution of ploughing at a time when the ravages of the rinderpest etc., amongst cattle and apprehension of its continuance mediate strongly against the policy of such a course and seem even to necessitate a reverse action". An arguement that was continued at some length.

In 1867 William Sellar and Henry Stephens collaborated in the production of a massive work of 634 pp. 8vo having the title *Physiology on the farm in aid of the rearing and feeding of lifestock.* It was copiously illustrated. These authors told their farming reader that the return on his livestock would be as a rule in proportion to the skill in management, a profound truism that hardly needed remarking (as so much of all that is written may be!). The book is what may be described as pseudo-scientific, but was excellent for that time. It describes the organs of nutrition in animals and is complete in all detail of its subject. It would be egregious to try to make a summary of its contents in this place, it being sufficient in my opinion to indicate its purpose, which the preface states very roughly in a not very original manner. Farming was followed, the authors say, for ages as an impractical art or not resting solely on experience, but was not generalised, but now (1867) begins to take a place among purists that claim to be rational in their character, which I suppose these men's textbook was intended to illustrate. This is not a cursory mention of what was I imagine an important and influential work but I have no copy immediately to hand. Perhaps I sold my library prematurely?

Irwin Edward Bainbridge Cox, a sonorous name indeed if not one to conjure with, wrote Vol. 3 of the Field Library, i.e., *The farm, garden stable and aviary* which first appeared in 1866, the second edition being produced 1869-1871, if that is not the date of the publications that made up the "Libary", G. H. Cox of London. This author does not seem to have been particularly interested in this special subject and his book requires nothing more than mention. He wrote a great many books, some half a dozen of which were about various hobbies, fishing, shooting and so on, the occupations of the leisured people of then rural England.

Accounts are important in every profession and no less for farmers than for everybody else. Consequently aspiring educators, or possibly it would be better to say authors, did not fail to set up as instructors in this difficult subject, perhaps rather more complex for farmers than for other professions where the interval between buying and sellingor making and selling is much more limited than the same period is for a farmer, whose mental outlook is prescribed by the seasons and whose seed time is several months before the resulting harvest. D.B., whose work is filed at Southampton tried to simplify this problem by producing *The agriculturists calculator; a series of tables for all engaged in agriculture or the management of landed property* an 8vo of no less than 591 pp. adorned with a frontispiece and issued by Blackie of London in 1867. No more need perhaps be said about this excellent piece of work — possibly because I have no more to say! It can be assessed as the little footprint daily washed away.

In the middle of the last century English breeding stock were being exported to such places as Cape Colony (as it was then called) and to the great ranches of the western states of America. One William Settle cmdr. of Stackhouse played a part in this business and wrote *The history of the rise and progress of the Killerby, Studley and Warlaby herd of Shorthorns* in an 8vo of 158 pp. decorated with a rather striking frontispiece, issued by W. Ridgway, London in 1869. Naturally these herds were not unknown; their history had for the most part already appeared in the form of letters in the pages of the Nash Lane Express and the Farmers' Magazine. The book opens with a description of the Booth Herds and goes on to make a point of Bakewell's work and that of Charles Colling of Ketton and Robert Colling of Barmpton, but a wide judjement of this book might be that it is genealogy, not the care, feeding and breeding of livestock.

In the same year, 1867, appeared *Practice with Science, A series of agricultural papers,* a 2 volume 8vo, Vol. I containing nine essays; Vol. II ten. It was published by Longmans. The contributors were as follows:
Vol. I 1. Rev. John Constable (q. v.) Agricultural Education considered in connexion with the R.A.S.E.

2. J. Bailey Denton (q.v.) Agricultural Drainage: the theory of underdraining as accepted by a practical man.
3. J. E. Ransome, Ploughs and Ploughing: a lecture dedicated to the R.A.C. (this is illustrated).
4. Arthur H. Church, Report on wheat experiments, 1863-64.
5. John Algernon Clarke, Farming and poultry: profits and management.
6. J. T. Harrison, Dairy farming.
7. Robert Warrington, Notes on the agricultural value of natural phosphates.
8. Rev John Constable, On the working of steam engines.
9. R. G. Welford, On leases (this writer became a judge!).

Vol. II 1. R. G. Welford (judge), Laws of real property as affecting agriculture.
2. John Wrightson, Professor of Agriculture at the Royal Agricultural College, Wheat experiments 1868.
3. John Wrightson, Experiments on barley.
4. John Algernon Clarke, Cattle grazing (at Long Sutton Lines).
5. Arthur J Hill, Farm Accounts.
6. Robert Warrington, On the absorptive power of soil.
7. John Wrightson, Experiments on clovers.
8. Y. H. Burges, The land question in Ireland economically considered.
9. W. T. Thistleton Dyer, The geological distribution of Ericalue Phosphate with a prefatory note by A. H. Church.
10. W. J. Edmonds, The feeding of stock.
11. Elias Pitts Squary, Agricultural Labour.
12. Charles L. Cantrell, Task work. Prefatory note by Rev. John Constable.
13. Rev John Constable, Rural education and the employment of women and children in agriculture (with illustrated plans of cottages).

Such a collection of crudition is difficult if not impossible to assess. These essays are of an instructive character, occasionally controversial and were mainly read at the Royal Agricultural College by speakers unconnected with the place or invited there for the purpose. The lecturers were mostly quite distinguished men of their time. The tables in the Task Work essay, it may be said, are instructive not only in their own right but because they provide a system of agricultural arithmetic that is, so far as I know, nowhere else to be found in relation to that date. The whole thing is a production that could be fabulously praised or viciously denigrated and the authors were so distinguished that they will perhaps appear again in the following pages.

There are many curious titles in the Perkins Library at Southampton. Though I once had some controversy with Perkins, I have nothing but admiration for his Library and his assiduity in collecting the scarcest and most recondite works great and small, some indeed very small.

One of them which is not to be found in the Bristol Library is by one William Peard, *Practical water farming*, an 8vo of 256 pp., published by Edmondston and Douglas of Edinburgh in 1868. I know nothing of the contents of this recondite work as it does not seem to be in the British Library.

I am trifle uncertain whether or not to include a 25 pp. 8vo of poems entitled *Cumberland Farm Life. Is a memorandum of old times Crossy Boggle*. It is signed D and the title appears in the Perkins Library Catalogue, which is all I know about it.

Robert Erskine Brown is quite a different kettle of fish. He was a factor and estate agent at Wass, Yorkshire and produced a large 8vo of 505 pp. generously illustrated over the title *The Book of the Landed Estate containing directions for the management and development of the resources of landed property.* W. Blackwood of Edinburgh and London published this rather monumental work in 1869. He was not I think a modest man but his preface proffers his excuse for producing the book: though there have been others on estate management, he had reason to believe that a book more suited to the advanced ideas and requirements of the present day than most of those which have hitherto appeared is not uncalled for (modest man).

The profit of the earth is for all, he justly observed. The king himself is served by the field. The book is certainly comprehensive — in a minor way encyclopaedic — dealing with Landowners, Agents and Factors. Farmers as a class, said Brown, were of great variety of character, men made up of many grades of society including such people as merchants and tradesmen, professional men of every description besides those born and brought up to the cultivation of the soil. Of such men, an example was Henry Stephens who had training on a farm and at an agricultural college.

His estimate was that the capital required was at least £10.00 an acre (shades of the past). Farms as an estate should not be small holdings. Home farms should be an example to the tenants of an estate. The whole production is garnished with simple axioms. All the details of farming labour, manuring, fencing (illustrated) the maintenance of buildings, land improvement (a generous term!), manures of all kinds, liquid and town sewage included, drainage, road making, the improvement of waste lands, traction engines versus horses, woods and plantations. There were plates of cottage plans and a widely different subject, the use of steam ploughs. There were advertisments of Stephes *Book of the farm* and Slight and Burn's *Book of farm implements and machines* as well as of other books.

The above was an omnipotent work. Brown also put out a print of his lecture *The cultivation of the soil by steam power, a lecture delivered before the Whitby Chamber of Agriculture,* an 8vo of 26 pp., adorned and implicated by 12 figures in the text. The illustrations are quite explicit and describe work with two engines and one, with the appropriate anchors, as well as giving extracts of costs and particulars of some more or less experimental work reclaiming waste and growing roots.

The disposal of town waste has always been a problem, possibly solved today but a subject of some complaint a century or more ago. A good deal was written offering solutions to what was definitely a difficulty at a time when town populations were increasing and much of the overcrowded cities being often little more than slums. Thomas Cargill was one who proffered a solution in his *Sewage and its general application to grass, cereal and root crops showing the results obtained by actul experience down to the present date,* a 52 pp. illustrated 8vo issued by Robertson Brown of London in 1869. Plans and sections illustrating the method of forming the ground for the different systems used and for distributing the sewage over irrigated fields. The diagrams are useful and informative and tables showing the results in different localities are supplied. Table III for example gives the number of floodings on land at Barking, its crop results and price obtained — all of which was intended to be convincing and certainly may have been to the interested and enquiring reader. I have no opinion at this late date.

A rather more imposing work appeared in bulk, and the impressive character of its comprehensive contents was *Rural Life described and illustrated in the management of horses, dogs, cows, sheep, pigs, and poultry . . . with authentic information on all that relates to modern farming, gardening, shooting, angling etc. And a complete system of modern veterinary practice,* a thick 4to of over one thousand pages, but only a trifle, i.e., 1016 pp, published by the London Printing and

Publishing Co. of London, New York. It was illustrated by upwards of 100 steel engraving which is numerically correct but to my taste deficient in artistic merit, indeed rather unattractive.

This is not true of the verbal contents although the writer (compiler!) was a man with a very good opinion of himself. It would always be a poor sort of man who had not.

The preface is composed with an almost hierarchical sense of observation from above to the humble reader and deserves extensive quotation of the well informed contents, which is not immediately possible, but I shall succumb to temptation and perhaps interest to make the citation Sherer deserves. In an air of humility Sherer wrote "Deferring to the opinion that the end of all writing ought to be instruction (some prefer entertainment!) but as instruction may belong to the past without, as well as within, the limits of existence allotted to men the object in Rural Life has been to approximate as nearly as possible to the present (not like the others e. g., *Old Norfolk Farmer* who proffered history as well as advice on curent problems) in order that while we are in the land of the living we may be participators in the enjoyment of the knowledge which, not only in Rural Life, but in every other condition is daily developing around us (the Victorian's belief in their achievements) and further progress" (where has it led us?). Veterinary advice is added.

The General Introduction despite his proclamation of up-to-dateness in the preface, discusses the classics, as I suppose was customary or accepted, because of the classical basis of education in those days, he purports. Among the ancients the literature of agriculture occupied a far higher place (distinction by attitude!) than that to which it has attained in modern times although equally there had been great advances in the past 20 years. He speaks of Mago and Hamilcar, repeating the accepted but rather doubtful legend about Mago. Sherer describes, or should I say criticises, these writers as non-progressive but repetitive of ancient wisdom. Sherer quotes Mago's directions for buying oxen for the plough but where this comes from, except at secondhand, I do not know.

The then modern veterinary practice was included by Sherer *in extenso*. Among other things this included, Cattle their varieties and diseases, sheep, pigs, poultry and finally Division VIII Principles and practice of modern English farming, illustrated by pictures of machines, still and at work. This comprehensive work is completed by sections on Horticulture, the Bee and the Apiary, Angling, all copiously illustrated.

No more need perhaps be said, except in admiration of the industry of such a man and the presumable accuracy of his description of the occupations of then current rural living.

I do not think I can omit an American publication, *How crops grow; treatise on the chemical composition, stucture etc. of the plant,* which claimed to be adapted for English use, which was revised with additions by A. H. Church & W. Thistleton Dyer, a 399 pp. illustrated 8vo, London 1869, but I do not propose to do more than mention it.

The first edition of *Cattle and the Cattle Breeders,* I have not been able to identify, but the second revised edition is a work of 135 pp., 8vo; it was published by W. Blackwood, London & Edinburgh in 1869. It is troublesome to write a simple note about a man like M'Combie who is described as the first Scottish farmer to gain a seat in the House of Commons and a man of long descent, the last an unoriginal comment because all men are of long descent. With more justice he was one of the greatest of the many great men whose lives have adorned Scottish agriculture, not all of whom had his distinguished political career. His father and grandfather were also dealers in cattle so he was born to be a breeder.

The essay on the feeding of cattle is mostly about his dealings but there is something about feeding on various substances. Essay II is Reminiscences and it may be said to be of considerable interest. III, Drovers and Robberies; IV, The cattle trade then and now; V, about Black Polled, Aberdeen and Angus cattle, and shorthorns. VI, Hints on the breeding and care of cattle (with some notes on diseases). It was authoritative and must have been widely read. A fourth edition edited by James Macdonald, editor of the *Livestock Journal,* was published by W. Blackwood & Sons of Edinburgh and London in 1886.

This decade might be concluded by a note about a man of some contemporary importance. He was R. O. Pringle, the editor of *The Farmer,* a periodical. He was a most industrious man (what ambitious man is not?) and wrote or published several quite important books at that time. His production of literature is generous but rather complicated. So far as I can resolve this problem I should say that his first attempt at fortune was *Purdon's Practical Farmer. The principles and practice of agriculture; including tillage farming and the management,* written by teachers employed at such places as Cirencester, Aspatria and so on where farming was taught by learned professors, or possibly supplemented as this seems to have been by the editor of a farming newspaper weekly or what have you!

Pringle also wrote *The livestock of the farm*, a copious work of 438 pp., 8vo, issued by Blackwood in 1874, a third edition of which came out in 1886 revised by James Macdonald. Pringle was a great noter of the obvious. He observed that the production of animal food was a matter of great importance and that the supply was then inadequate to meet the demand so that the value of meat increased and it became a luxury for the poor. In this respect there was ample room for improvement. The repeal of the Corn Laws had not had the anticipated results of large imports from the continent while home bred animals had been destroyed by the outbreaks of contagious disease which he attributed to imported cattle. He supplied tables showing the amounts of different substances that might be fed with advantage to the animal and to the breeder whose beasts improved. An appendix is a chart for weight, girth and the length of cattle.

A commendable piece, or pieces, of writing with which to conclude the farming textbooks and treatises of the decade.

III

1870-1900

The increase in the number of teaching institutions of whatever sort necessarily leads to the increased production of textbooks. As agricultural educational institutions began to increase the number of professors bore a proportional relation because somebody has to teach. The best thing was for these purists to write textbooks which they could supply to the students. They had an established and continuing public in the succeeding generations of students who followed in an agreeable and profitable succession, as did the textbooks by their teachers which the students had to buy in order to know what answers to give to the questions posed at their examinations which, if success greeted their efforts, led to remunerative jobs in the countryside — not so numerous or official as today — but sufficient to give the lettered student a chance of living.

There were a good many professors who, reasonably enough, took advantage of their positions to write textbooks of farming. It may be unfair and dubiously critical to say that the authors and publishers of such works did not aim at the practicing farmer but at the college student. Of such was R. O. Pringle and his collaborators whose works are discussed in the previous chapter.

William Brown who wrote *British Sheep Farming,* an illustrated 8vo of 129 pp., is alphabetically the first of the farming writers whose work came out in the 1870's. Brown's book was indeed printed in 1870 by A. & C. Black of Edinburgh.

Next he seems to have produced a 2 vol. work of a co-operative character over the title *Practical farming in a series of treatises by R. O. Pringle, Prof Murray and other distinguished agriculturalists.* This was an 8vo published by A. Fullarton, Edinburgh & London. It was adorned with many illustrations both figures and plates but was a rather curiously planned book as a set of five treatises, the authors of which were all eminent in writing if not in farming: the whole makes a sort of encyclopaedia of reference. Victorian authors sometimes adopted odd arrangements of their matter. This is one such book but it has a very excellent system based upon the analysis of various subjects. The titles of the five departments are (1) The Cow, (2) The Dairy, (3) The management of fattening cattle, (4) Diseases of stock on the farm (by A. J. Murray Professor of veterinary medicine at Cirencester), (5) Horses, sheep, pigs and poultry (by the author of *Notes on cattle, food and feeding* and the author of the *Poultry Kalendar*.

The preface says the book was the first of the series *Home and Farm Miscellany* and that it was based on the celebrated treatises of M. M. Magne and M. A. Thier. Magne's *Chois des vaches laitières* was, said Pringle, accepted on the continent as of the first importance and an improvement on Guenon. Thier, *La laiterie* is mentioned and the editor of this volume admits that he selected from these works what seemed to him best adapted to English practice. He also used other continental writers, indeed the whole thing is derivative though an amalgam of his authorities. It seems to have been one of the increasing number of textbooks. This has a coloured map of the British Isles as a frontispiece "exhibiting" the distribution of various kinds of sheep in Great Britain and at the end a coloured diagram displaying the "characteristics" connected with sheep in Britain. In his preface Brown makes the commonplace criticisms of his predecessors and of his contemporaries. He was fond of statistics and reprints some tables from the almost novel *Agricultural Statistics,* and discusses the physical geography of the whole country as well as the ditribution of the different breeds of sheep, as well as the products in wool and mutton − a curious and rather fascinating book upon which I could expound largely but I think I have said enough for the purposes of this work and its possible readers.

William Henry Corfield returned to the ever present problem of the disposal and use of sewage which had absorbed the attention of the British Association 1869-70. Corfield prepared for that meeting *A digest of facts relating to the treatment and utilisation of sewage,* which was published by Macmillan in 1870 as an 8vo of 282pp. The British Association had considered it desirable that a digest of all matters relating to the above subject should be prepared and this became Corfield's job. His work was divided into two parts, the Treatment of Sewage and the Utilization of Sewage. The major part of the book deals with the disposal of sewage and is therefore of little or no account in relation to agriculture except in rural areas where perhaps it may have been useful.

A couple of years later (indeed the subject was perennial!) Ralph Ulick Burke, a barrister, returned to the subject with *Sewage Utilization,* a handbook of a small 8vo of 60pp published by Spon, London 1872. A second edition of this work was put out both in London and New York but this work has no rural reference and perhaps should not have been included here.

Sir William Crookes, whose fame in other connections is well enough known or ought to be, was a scientist with a productive pen. For example he wrote one 290 pp. memo, 8vo, *On the manufacture of beet root sugar in England and Ireland* with an appendix on Baumes Arcometer, but this is not about growing sugar beet but the process of making sugar.

Crookes translated and edited Georges Ville's work and published it as *On artificial manures; their chemical selection and scientific application to agriculture. A series of lectures given at the experimental farm at Vincennes during 1867 and 1874-5,* published as an illustrated 8vo of 450pp by Longmans. There were 31 engravings and a frontispiece of diagrams e.g., to quote The power of production of the chief systems of cultivation in their historical order, supposed to be from before classical times, but is in that respect not particulary satisfactory.

The preface to the English edition is an exordium on the then present agricultural crisis in England and France which he thought was only the prelude to an economic struggle between the Old World bound in the old tradition and the New World pressing forward unrestrainedly in the path of progress, a piece of rhodomontade that I find rather pleasing.

It was, he said, an undeniable fact (can fact be denied?) that except under rare and altogether exceptional circumstances, farming operations carried on solely with manure produced on the farm itself have for a long time ceased to be remunerative. To obtain certain profit one must have recourse to manufactured manures.

The book comprised a series of 15 lectures which could be largely elaborated but about which I will try to be succinct, voluminous as these were. Naturally the contents were of a very exhaustive and technical character covering such ground as Theory and practice, the composition of plants and their absorption of manure in the shape of chemicals, typical fertilisers, the comparative cost of farmyard and chemical manure coupled with the assertion the artificials must be used. The waste of crops were effective manure, e.g., haulm and pods of colza. Past and present systems of agriculture are touched upon — irrigation, cattle grazing, the triennial system, the quantity of farmyard manure required for similar crops, systems of alternate rotation and much more. This is all very technical, perhaps too much so for the average farmer, but perhaps the students of agricultural colleges and so on could have used this doubtless valuable matter.

Much later Crookes put out his *The wheat problem. Based on remarks made in the Presidential Address to the British Association at Bristol in 1898*. This was revised and supplied with two chapters on the future wheat supply of the United States by C. W. Davis and J. Hyde, etc., an 8vo of 207 pp. published by Murray in 1899: but a great deal has been written about Crookes, and his work need not be further expanded here, as much of it was scientific and political rather than agricultural.

Hozier is a well known name, or at least it was in the 1920's and 1930's. A farmer of that name wrote "A gold medal winning report to the Highland and Agricultural Society published at their direction". Its title was *Practical remarks on agricultural drainage*, an Edinburgh 8vo of 1870. It was "Especially adapted to the drainage of heavy land with some observations on subsequent management", some 76 pp. Hozier profoundly remarked that in order to develop a perfect growth of crops a fine circulation of air and water must be maintained throughout the soil, and emphasises the harm of saturated soil in an expanded discussion of the subject. His catalogue of the necessary equipment is comprehensive and is illustrated by pictures of pipes, levelling instrument and so on. He gave credit where credit was due, saying that since the days of Mr Smith of Deanston the system of drainage invented by him has been universally adopted in Britain, praise indeed of a disciple but a rather exaggerated statement, though having some foundation. The process need not be set out here. I have discussed it elsewhere.

There was at this date, as there had often been before and was repeated at frequent later dates, a great deal of discussion of the value — to society at large — of small farms and small holding, which could be undertaken and successfully worked by men of small capital but of sufficient energy and ingenuity. One of the protagonists of this idea was a Miss Coulton who published a description of *Our farm of four acres and the money we made by it*, a 118 pp. 8vo over the inprint of Chapman and Hall in 1859. The book has a nice frontispiece depicting a house, or should I say the house, with some children shewn in the foreground. The author opens chapter I with the obvious question, where shall we live? It was essential that they keep a cow so they took the obvious step and bought one for milking and making butter. A little later they got a second one and from the two they made a profit of £15.18.4 (commendable precision!). They kept pigs and feed poulty and pigeons. They cured ham and made bread. They laid out a kitchen garden, the profits of which enabled them to keep a pony. The book was popular, a tenth edition revised and enlarged, an illustrated 8vo of 208 pp. being put out by the same publishers in 1870. It was a justly popular book, as the many editions demonstrate, said a comtemporary, but successful as it was, not much more than a pleasant and profitable hobby. This author put out a farming calendar (of 239 pp.) over the title *From hay time to hopping,* issued by Chapman and Hall in 1860. Its title indicates the contents of the book and it needs no further description, its contents being self suggestion.

The Perkins Library has *British Sheep Farming,* an illustrated 8vo of 129 pp. published by A. & C. Black of Edinburgh, but this does not appear to be in the British Library.

Sir John Bennett Lewes, first Baronet of that ilk, bears a name that is sufficiently well known to the farming community — perhaps today but certainly in his own time and the following decades — through his very well known army of students and others interested in the progress of agriculturual science, its development and use. As well as making practical experiments on the soil of Rothamstead and on other farms, he wrote voluminously and continuously as others, afflicted the *cacoethes sinbendi* are forced to do by their nature and ability.

Whether I have been able to make a complete list of his writings is a question I dare not ask and certainly cannot answer. The first of this long list seems to be *Exhaustion of the soil in relation to landlord's covenants and the valuation of unexhausted improvements,* an 8vo of 51 pp. issued by Rogerson and Tuxford, London 1870. Its title is sufficiently explanatory of its contents. The subject was one of consequence to both landlord and tenant, incoming or outgoing both.

Sir John made many contributions to periodicals of one sort and another, some learned, some vulgar, and it would not be profitable to attempt to discuss them all in this place. His next small thing was *On the more frequent growth of barley on heavy land,* a 20 pp.

pamphlet printed by W. Clowes, London 1875, a brief recapitulation of his theories and ideas on the subject. The Journal of the Royal Agricultural Society published many of Sir John's essays but I think I should confine myself here to what I have already said.

A very different kind of writer was Alfred Stables of Kirkbank Richmond Yorkshire. He wrote a 32 pp. 8vo pamphlet on *The cultivation of land by steam power; an essay for which Lord Wenlock's prize of £10.00.00 was awarded by the Wenlock Farmer's Club May 1870.* It opens with a discussion of the advantages of steam cultivation, and of the use of the appliance as an economic agent. Some costings are supplied. Stables protests that in making the statements contained in this essay "I am biased by no motive save that of a wish to promote the interests of agriculture," and indeed at that date steam cultivation was very much in the interests of agriculture, especially those of the farmers who worked a large area, indeed a pungent argument.

Whether one of Purdons' Veterinary Handbooks ought to be included here is an open question but R. O. Pringle wrote *Diseases of horses, cattle, sheep, swine, dogs etc.*, a second edition of which came out in 1871. This author gives a lengthy list of authorities including Youatt, Fitzwilliam etc. Pringle protests in his introduction that his object was to afford information relative to diseases of domesticated animals for those not being within reach of professional advice in cases of emergency. "First we are anxious to express advice that most of these are capable of prevention by forethought and ordinary care". A list of authorities consulted is given and a description of each type of animal as then living is supplied in some detail, which is an appropriate reason for the inclusion of the book here. It must have been of great use to farmers of that day. Its detailed descriptions and advice can be clearly understood.

Sheep have always, at least during historic times, been important in human economy, supplying wool, meat, manure from their grazing on the fallow field in the days of open field cultivation to their fold on the enclosed fields of later days. Consequently William Reid's book about them was important to the agricultural community of his day and is still very much so as an historical source book. Its title is *Sheep; their history, management, diseases and national value; with remarks on the transit of stock,* an illustrated 8vo of 146 pp. put out by W. P. Normans and A. Elliott, Edinburgh in 1871.

In his foreface Reid mourns the decline of Great Britain as a wool growing country, since it was then rapidly costing its position, a misfortune since the current expansion of population required the opposite tendency. Reid suplied the history of the animal, noticed their early mention in scripture, discusses wild sheep and the varities of this useful animal as well as the perennial subject of Italian and Spanish Merinos.

Reid was an extremely knowledgeable man and well acquainted with the details of his subject. He describes each breed of British sheep, the Leceister, Lincoln, Cotswold, Romney Marsh, and Black faced. In his second chapter, management, choice, clipping, foot and mouth disease are discussed, while turnips are recommended as a cure for brasey.

His interest in the history of his subject, or should I say subjects, leads him to describe the transit of stock both before and after railways, a full treatment of his subject and a fascinating book. The end of the book is a galaxy of advertisements, modestly including some of his own.

It may be thought a trifle odd to include in these pages the Rev James Stormonth's *Etymological and Pronouncing Dictionary of the English Language,* a large 8vo of some 753 pp. published by Wm. Blackwood of Edinburgh. It is in very small print, a frequent but characteristic fault in Victorian publications. So great a capacity may safely be described as exhaustion. Naturally a dictionary includes many subjects other than farming, but the subject could not possibly be avoided in such a work then, before 1871 and even today. An example of Stormonth's condition, or possibly of his researches, is the word Plough and its connections, which is dealt with quite exhaustively. Of course the compilation is not an agricultural work, but it abounds with farming words, their definition and so on, and for this reason I feel quite justified in including it here; nor do I excuse it.

The disposal of sewage could not possibly be a 'burning' question but it was certainly a problem in the growing towns of Victorian Great Britain. Mick Ralph Burke, Barrister at Law, was one of many who discussed the question in the hope of disposing of the accumulation of this material, unwanted, difficult and even dangerous to health if allowed to accumulate when it was produced, as indeed it was in some of the lower cellars of some great mansions.

Burke's solution of the problem is defined in a 60 pp. 8vo, *A handbook of sewage utilisation* printed by E. and F. Spon in 1872, both in London and New York. A second edition came out in 1873. On the disposal of sewage Burke says he took up the subject without any prejudices or any preconceived notions. "I have nothing to support..." There were two angles to the problem (1) pollution (2) systematic waste on a gigantic scale of a valuable manure (which was also a source of pollution of rivers to an almost incredible scale). Liquid sewage should be used on the appropriate crops (he does not say what they were!). His theories and ideas could be expanded but enough has been said of him and his pamphlet at this time.

Another 'burning' question in the 1870's was that of the losses incurred by the infectious and widespread disease in the potato crop. Several writers exercised their eloquence on the subject, amongst them one James Craig, who wrote what he was pleased to call *A common sense view of the potato disease showing how it comes and how to prevent it,* a 442 pp. 8vo, put out by Longmans, London and Johnson and Tissyman of York in 1872. The printed cover says that Craig was a landscape gardener of Hepworth, York. The little book is dedicated to the Royal Agricultural Society.

Craig was by no means a modest man. He felt it was his duty to bring before you (the R.A.S.E.) and the public at large, without longer delay,"such facts and such reservations as have arisen therefore or have recurred to me during a somewhat lengthened practical experience. One peculiarity in the cultivation of the potato which in my opinion is far too little thought about . . . unlike all our field crops which are invariably propagated from seed, the potato is promulgated almost entirely from cuttings or off shoots and raising it from seed being quite an exception". He follows with a discussion of types, yet he advertises "A packet of potato seeds saved from the healthiest and best . . ." and he offered to send a packet of seeds free by post on receipt of sixpence (in stamps). He also offered Craig's prolific early cucumber seeds post free for 13 stamps per packet. Certain it is that some of the Victorian farming propagandists were very engaging characters.

Another writer on this delecatable subject was Frederick Hahn Danchell, whose pamphlet was one of 41 pp., 8vo, published by Simpkin Marshall in 1872; but this is really more about the clearance of town wastes than their agricultural use. He was cynical about the protagonists of the subject. "To attend a meeting of sewage doctors, and hear them wrangle", he wrote, "and implode each others doctrines is enough to bewilder any unscientific auditor. Really little help is to be had from authority and a plain man must be content to trust his own judgement", a severe but sensible criticism.

Finally for the year 1872 there is *Clodhopper cracks vegetable and animal manufacture* by David Curror, a Fife Scot, some 128 pp., 8vo, issued by Seton and Mackenzie, Edinburgh 1872, but I could not find this in the British Library

A book rather in the nature of a popular condensation of a subject which had a long and somewhat varied life, was an anonymous production, one of Warne's Country Library: *The sheep, its varieties and management in health and disease. Condensed from Morton's Farmers Calendar, Manuralia* etc., this was an 8vo of 119 pp., adormed with a frontispiece and other illustrations, dated 1873. The development and editions of this like others of Warne's productions are both difficult to follow and to distinguish. A new edition, revised and enlarged, was issued in 1852. Other editions seem to have been printed in 1893. I am a trifle uncertain about this work, stated to have been revised and enlarged by George Armitage in an edition of 1894, but nothing is presented in the publication itself to suggest that Armitage is its author. I do not think it necessary to summarize its contents here, as they are of an orthodox and usual kind.

One John P. Gould gave a lecture, afterwards printed, as so many of these fugitive pieces were, on *The chemistry of agriculture* to the Members of the Croydon Farmers' Club (shades of a century and more ago! There are few if any farmers in Croydon today). The lecture was printed as a 30 pp. 8vo by H. M. Pollett of London in 1874. No copy of this immortal work is filed in the British Library, so all I can do is to record its one time existence.

A civil engineer whose repute gained him a note in the D.N.B. was named David Stevenson. He published an 8vo *On the reclamation and protection of agriculturual land,* which was printed at Edinburgh in 1874. His biography states that he became a recognised authority on the improvement of rivers and estuaries, docks and so on but such work is only minimally agricultural, and I think that a cursory reference to this man, undoubtedly of consquence in his particular craft, is supplied here. He spent some of his youth in the workshops of millwrights and wrote a great deal, for example for the *Encyclopaedia Britannica* and so on. No more need be said of him in these pages.

Frederick Clifford's *The agriculture lockout of 1874 with Notes upon Farming and Farm Labour in the Eastern Counties,* a London 8vo of 1875 must perhaps be mentioned, but I tremble at the thought of discussing union policy, so will merely mention this undoubtedly valuable record.

John Coleman was of a very different type. He wrote *The cattle of Great Britain, A series of articles on the various breeds of the United Kingdom, their history and management,* a substantial volume of 151 pp. The preface states that "A series of articles on the breeds of British cattle appeared in the field during 1872-1873, supplied by eminent authorities. These are here presented with the original illustrations of Mr Harrison Weir, mainly sketched from life and here copied from photographs or reliable drawings. Copious notes and additions are added to bring them up to date. Anyone travelling through the country must notice the progress in the general quality of cattle. Buying a bull with a pedigree and not enquiring into the milking qualities of its family background is not wise. In these days when it is the fashion to rear calves, well bred country calves sell at remunerative prices. We must impress upon farmers generally the importance attached to colour in choosing animals to breed from, so called colourly stock finding a much readier sale than when the predominant colour is white. Graziers do not like them, will not have them if possible, thought not to be so hardy as reds and roans. It is a fact that white animals seem more liable to parasites than their richer coloured companions. Why we cannot vouch but there can be no doubt as to its truth".

Breeding and careful management is dealt with at length plus food and its nature, buildings and manure, dairy management and the milk trade. In Part II each breed is described in detail, its size and shape. The illustrations are excellent and help the reader to visualize the animals discussed.

One William Lawson and others co-operated in producing *Ten years gentleman farming at Blennerhasset with co-operative objects* only the 2nd edition of which we have seen. It was published in 1875 and was an 8vo of 408 pp. illustrated with plates over the imprint of Longmans Green of London. The purpose of this book, says the preface, can only be fully understood by the perusal of it, another glimpse of the obvious. It is illustrated by tinted plans, contains many tables of one sort and another, and some descriptions of experiments.

An introduction is supplied by G. J. Holyoak. "The reader is informed", he wrote, "that he will soon realize that Mr Lawson never hesitated to tell a story against himself, writing with self deprecating candour as if he does not actually take pleasure in giving the reader an impression to his own disadvantage, he takes no trouble to make a favourable one". The reader will find a number of curious local incidents and proceedings.

Mr Lawson left his father's home to ride to London, taking thirty three days on the road visiting friends and places of interest. He was ready to talk to anyone, thus acquiring a good deal of local knowledge which changed his former impression that farming was dull, but he was much impressed by Alderman Riche's experiments, and when he returned home tried to impress his father (a difficult task for a farmer's son) but must have done so when in October 1861 he was offered Blennerhassett Farm to experiment on, if he cared to take it, which he obviously did. Lawson stimulated his workers by giving a bonus when he thought it deserved.

There is some description of the primitive methods of horse drawn transport, and travelling in farm carts — a very luxurious process. Some family accounts, tables and methods are supplied. The general impression arrived from a perusal of this work is one of a forward looking man not destitute of ideas, a man of great interest today because of his work, and doubtless of influence in his own lifetime.

During the second half of the 19th century there was a plethora of political writing about the condition of the farm worker, both by members of his own class and by his own class, but this is not the place to discuss this sort of polemic: this work purports to be about farming textbooks and I have dealt with the subject in what is now doubtless a long forgotten book, *Tolpuddle to T.U.C.*; but I think I may be permitted to mention, if no more, George Mitchell, *The skeleton at the plough; or the poor farm labourers of the west and with the autobiography and reminiscences of George Mitchell, One from the ploughed* by Stephen Price, an 8vo of 158 pp. published by G. Potter, London 1874. I have no valid excuses for doing this except that this protest intrigued me.

A much more precisely agricultural work was written by Patrick Sherriff. It was *Improvement of the cereals and an essay on the wheat fly,* an 8vo with a frontispiece, a portrait of the author and other plates, printed for private circulation over the inprint of Wm. Blackwood of Edinburgh and London in 1873. The illustrations I shows part of the plant, and II wheat. The book was well received and it won a medium gold medal of the Highland and Agricultural Society.

Many of the 19th century writers of books about farming were apparently afraid of the lack of interest that would be shown in their work and make profound statements in their preface about the pressure that had been put upon them to write the book. Sherriff is no exception to their rather arbitrarily pronouncement. "Having frequently been asked," he wrote "to accept of a testimonial for having raised and introduced new varieties of grain this high compliment was at first declined on account of my advanced age but at last I acceded to the wish of my friends on condition that the money raised for the testimonial should be set aside for improving the cereals in time to come . . . I therefore resolved to prepare a short account of my experiences while engaged in improving the cereals." That he was a man of personal credit is demonstrated by the list of subscribers which was headed by the Marquess of Tweeddale and included many other famous men. This is a sufficient appreciation of the value of the work, the contents of which need not be expanded here as the title and so on makes them self evident.

Sherriff also wrote a *Tour through North America* which describes the farming he saw in Canada and the United States. It was an Edinburgh 8vo published by Muir and Boyd of as so long before as 1835. This perhaps only needs mention here.

This is not the same of John Usher's book *Border breeds of sheep. With an appendix.* It is a square 12mo and was printed by J. and J. H. Rutherford of Kelso in 1875. This small production was based on articles that had appeared in the field, but properly revised, and deals with Blackfaced sheep, Cheviot sheep and Border Leicesters. Usher praises his fellow breeders and thought that the Blackfaced breed had been improved by men of intelligence and judgement in careful selection. Usher thought the Cheviots were a native breed, though then was a legend that they swam ashore from wrecked vessels. The Border Leicester was, he thought, derived from Bakewell's stock, a theory that is supported by later writers. The book is completed by an appendix about a famous breeder of Cheviots, one William Aitchison of Briery Hill, a place I have been unable to identify, so where it is or was I do not know.

The Perkins Library has two pamphlets by Thomas Jamieson which are not to be found (by me at least) in the British Library. These are *The relation of chemistry to agriculture; a lecture delivered* at *Aberdeen* (and elsewhere), an 8vo with 24 plates put out by the Aberdeen Journal office in 1876 and the *Opening lecture, Aberdeen School of Chemistry and agriculture, season 1878-9. Subject, investigations conducted by the Aberdeenshire Agricultural Association* an 8vo of 28 pp. and a plate, issued by A. Murray, Aberdeen in 1879. Another thing by this writer is *Contribution to the land question, Compensation for manures,* printed by J. A. Churchill of London in 1883. I do not need to expand the contents of this production.

The Statistical Society of London founded what was known as the Howard medal to be awarded for the Howard Prize Essay in 1875. This was only Edward Smith in 1876 with an essay *The Peasants' Home 1760-1875* which was printed by Stanford of London in 8vo, 1876. It ran to 139pp. It professed to supply a general view of the efforts which have been made since about the middle of the 18th century to raise the condition of the British peasant by the improvement of his home.

It describes the commonplace, even usual, bad conditions. Some housing was in farm houses that had been abandoned. "Many timber and lath and plaster tenements were," he wrote, "sounder in frame than some of the 'model' cottages of today (1876)." Much of this book is based upon the Royal Commission of 1867 on the Employment of Women and Children in Agriculture. Somerset and Devon were, said Smith, probably the worst counties in England. The balance of the book seems to be a general survey, county by county, of the then present conditions, but with some backlash in references to the County Reports. Rural housing was then in a deplorable state in many parts of the country though some model cottages were built here and there by philanthropic, not to say compassionate, landlords (if there were such!)

In the early years of the last quarter of the 19th century there was a good deal of discussion of the preservation of forage crops as ensilage. Prominent amongst those who entered these list was one Thomas Christy, Jnr, F.L.S. He wrote *Forage plants and their economic conservation by the new system of ensilage,* an illustrated 8vo of 72pp., published by Christy and Co. of London in 1877. The same firm also printed *Ensilage, a system for the preservation in pits of forage plants and grasses independent of the weather . . .* a 6pp. 8vo of 1883.

Part I of the first book dealt with Caucasian Prickly Comfrey: Part II with Balsamucarpoe and the plants containing tannin. There is a coloured frontispiece to this showing various cuttings and so on. The preface describes the pamphlet as a popular exposition of the qualities of certain forage plants and shrubs to which is added an exploitation of the merits of certain plant machinery essential to the system of *Ensilage* or picking fodder. In this commission Christy refers to one Mons. Barrell, the distinguished French Professor of Agriculture. England, Christy wrote, has splendid land for forage crops which was a great advantage, as the stock that fed upon then always commanded high prices. Comfrey, he pronounced, is wonderful.

Christy then proceeded to other things, such as the difficulty of getting farm servants. "The progress of education", he declared,' 'is rapidly making labouring people feel above their natured employments and the rise in wages (if any) makes it incumbent upon scientific farmers to diminish the amount of labour by the largest possible use of machinery", illustrations of some machines being supplied such as a portable steam engine, a self-acting manure fork, greenfodder and chaff cutters.

Most of the Victorian writers followed the fashion of their day (who doesn't) and provided elaborate and explanatory titles to their writings. They certainly did not believe that gravity was the sand of will! John Coleman, for example, described his work by the title *The sheep and pigs of Great Britain. A series of articles on the various breeds of sheep and pigs of the United Kingdom. Their history and management.* It was an illustrated 8vo of 206 pp. and was put out by H. Cox of the *Field*. He was nothing if not obvious, proclaiming in his introduction that sheep are prominent in our history of agriculture. To them our progenitors were indebted for much national prosperity. Dionysius Alexandrinus tells us that "The wool of Great Britain is often spun so fine that it is a manner comparable to the spiders web . . . The Lord Channellor sits on the wool sack and maidens of all degrees were taught to spin, hence the term 'spinsters' ".

The management of ewes up to lambing is discussed, and from birth to bearing, and also the wools the animals carried. The book contains an essay on each breed by an author of repute and illustrations that must have been then, and still are, instruction for the enquiring reader.

The Perkins Library has a pamphlet by Alexander Harvey, *The cereals; their natural history; and the testimony they bear to the supernatural in nature* . . . issued by the Aberdeen Journal office, a 42 pp., 8vo, but this we could not trace in the British Library. That collection has *On a remarkable effect of cross breeding*, some 39 pp. 8vo, issued by W. Blackwood, of Edinburgh and London, which seems to have been put out some twenty-five years before in Aberdeen, 1851, but it is not clear whether this is by the same author. There is a name, Dr. Dran Tarland, on the title page but I cannot claim to have seen it.

Hops have been important for several centuries, since the 16th at least, and Peter Land Simmonds added his quote to the literature of the subject with *Hops, their cultivation, commerce and uses in various countries. A manual of reference for the grower, dealer and brewer*, an 8vo of 135 pp. issued by Earl F. Spon of London and New York in 1877. The preface claims a long connection with agricultural literature and that an intensive home and foreign and colonial correspondance with cultivators and agricultural societies "have shown me how rapidly hop cultivation is extending with the increased demand for beer." Modestly this man "does not hope to afford much practical or novel information," but only to provide a few useful hints and suggestions, as well as the latest statistics of production and consumption.

This author, who was apparently a professional general writer, also produced *Waste products and undeveloped substances; a synopsis of progress made in their economic utilisation during the last quarter of a century at home and abroad,* only the third edition of which we have seen, an 8vo of 491 pp. issued by Hardwicke and Bogue, London in 1876.

Simmonds also wrote one of the South Kensington Museum Science Handbooks, *Animal products, their preparaion, commercial uses and value*, a 416 pp. illustrated 8vo put out by Chapman and Hall London in 1877. This was followed by an economic rather than agricultural work *The animal food resources of different nations*, a London 8vo of 1855. I think Simmonds was a professional writer, perhaps a farming expert but perhaps only an assiduous reader and writer, with which equivocal comment I will close any further comment upon this individual.

John Wrightson was professor of Agriculture at the Royal Agriculturual College at Cirencester and afterwards at the similar establishment at Downton. Men in this sort of job always write books, textbooks and so on, for which they have a ready made market in the pupils they instruct. Wrightson was no exception to this rule, if rule it can be called. He wrote many books all of which I hope to have included here, but I must cringe with shame if I have omitted any of these prolific writers' works — a very marked possibility. I suffer from human failure as much as the next bibliographer.

The first of Wrightson's productions was a textbook entitled *Agricultural Textbook embracing soils manures, rotation of crops and lives adapted to the requirements of the syllabus of the Science and Art Department, South Kensington.* This was one of Collins' Elementary Science series, 208 pp. with a frontispiece and other illustrations published by W. Collins of London and Glasgow in 1877, another edition of which was put out ten years later. It is not necessary to expand on the contents of this work, which are sufficiently obvious from the title, as indeed are those of *Fallow and Fodder Crops* which Chapman and Hall published in 1889. Cassell's Agricultural Readers, Downlon Series, included a title *Farm crops* an illustrated 8vo of 224 pp., 1891. Before that *The principles of agricultural practice as an instructional subject* had been put out by the same firm in 1888, a 218 pp. 8vo with a frontispiece and other illustrations, a 3rd edition of which came out in 1893 ten pages longer, 228 pp. 8vo.

Sheep, breeds and management, one of Vinton's Live Stock Handbooks, was published in 1908, a 235 pp. work, an 8vo with plates; but this writer's works went on into the twentieth century, a long list; but in 1892 a 200 pp. 8vo on *Livestock* was included in Cassell's Agricultural Readers. Wrightson also wrote articles for the periodical *Land,* and these, or some, were reprinted as *Technical instruction in agriculture,* a 24 pp. 8vo with a questionable or rather uncertain date of 1882. Another lengthy work of 648 pp. called *Agriculture theoretical and practical* was reprinted in 1919 but that is well outside the period dealt with here, though publications by particular writers are impossible to confine within certain decades. This went into a 3rd edition in 1921. I fear this prolific writer and professional teacher lived and worked both within and beyond my limit of 1900. He was a teacher and professor, and, as I have said, his numerous works must be treated with respect and perhaps as a reflection of the common practice of his day.

In the mid and towards the end of the 19th century farms were parts of great estates, estates owned by the great landowners of that time. Consquently a book about *Estate Management. A practical handbook for Landlords, stewards and pupils. With a legal supplement by a barrister (Frederick Green). Also Tenant Rights from a landlord's point of view* is properly included here. It was an illustrated 8vo of 332 pp., issued by H. Cox, the Field, London in 1879. A second edition was published in 1882 and a fourth expanded to 428 pp. in 1898. The author was Charles Edward Curtis.

Curtis was certainly a man of judgement, as was of course to be expected of a man of his profession. In his preface he wrote "The message of knowledge for good conduct of an estate (is) generally admitted to be that a complete and thorough course of study must be undertaken". The book was not aimed at exhaustive and extension discussion but was just an outline of necessary matters. Although this was the author's principal objective, he hoped the work might be of some service to the profession generally. The contents fulfilled his promise, including all aspects of his subject beginning with the legal side, compensation for unexhausted improvements, general valuation etc; all very detailed and plain, adapted to every intelligence. The book must have been a valuable asset for those who had it for permanent reference.

Curtis wrote on other subjects not of immediate interest here, *Practical Forestry*, for example, *The Valuation of land and houses* and other works on *Valuation* as well as *Farm buildings for landowners, agents and tenants,* an illustrated work of 144 pp. issued by Vinton and Co., London.

A pamphleteer, one James Pink wrote on that constantly recurring subject, *Potatoes, how to grow and show them. A practical guide to the cultivation and general treatment of the potato,* only the second edition of which is filed in the British Library. It was an illustrated 8vo of 95 pp., put out by C. Lockwood, London in 1879. "The value of the potato and the interest taken in its cultivation at the present day must be my excuse for laying before the public my method of cultivation". Pink asserted that he had received the greatest number of prizes of any one exhibitor at the great International Potato Exhibition. He added that there was no royal road to success (is there in any profession?), the only secret being "Good Culture, the principles of which I have endeavoured to give in the following pages". His first chapter disclosed "The relative value as an article of food". Pink speculates about the introduction of the plant and who was responsible and he cites Johnston, *Elements of Agricultural Chemistry* as one of his authorities. Pink also wrote *Potatoes or how to grow one thousand pounds of potatoes from one pound of seed*, an optimistic but not impossible undertaking, but like that of so many others Pink's effusion need not be laboured.

In the same year 1879 Alxander Ramsay produced his *History of the Highland and Agricultural Society of Scotland*, a 592 pp. crown 8vo published by Blackwood and Son, Edinburgh and London. This has an introductory chapter dealing with "Anterior Societies" e.g. the Society of Inprovers, the Edinburgh Society and so on, but this subject need not be laboured here. A similar mention of this work is a sufficient record.

In one of Chapman and Hall's catalogues, a work by Arthur Rowland, *Farming for pleasure and profit. Dairy farming, management of cows etc.*, edited by William Ablett, a large crown 8vo, is advertised, but this we could not find in the British Library.
The average farmer did not like white cows but insisted that none of this negative sort of colour should have a place in his herds; all the same the wild white cattle of the north were famous and renowned so much so towards the end of the 19th century that one John Storer was stimulated to write an 8vo of some 384 pp. adorned with some charming illustrations entitled *The wild white cattle of Great Britain. An account of their origin, history and present state* which was published by Cassell, London. It was unfortunate, as the preface states, that the Rev J. Storer did not live to altogether complete his researches. With the exception of the Hamilton and Kilmoney herds, the account may be regarded as substantially as he would have published them, except for last verbal corrections. The earliest portion of the book contains a complete general history of the wild cattle of that country and kindred races abroad.

The introduction states that some herds were extinct. Storer wished to call attention to these most ancient races preserved in Great Britain alone. At the time of writing these white herds were reduced according to his account to five. It seems remarkable that such races survived so long, for their colour is distilled by most British breeds. In the remote parts of Sussex, Devon, Wales and the Highlands of Scotland, the colour white has been the fashion for ages. In Ireland, as in many other parts in the present day, a white bull is nearly or quite unsaleable. The book opens with a detailed account of the history and origin of cattle, unto the Chillingham herd, undoubtedly one of the finest of wild white herds.

Later in the book he wrote "The ancient predilection for white cattle was a curious coincidence . . . Most valuable as sacrifical offerings to the gods which was a practice of the Romans in Britain". The book is written with great care and detail and much history is included with that of the breeding and use of this sort of cattle.

One A. W. Crews wrote *Manures; their respective merits from an economical point of view,* a small 8vo of 133 pp. from the Field Office, Horace Cox, 1880, but this does not seem to be filed in the British Library. Crews was a considerate man who wanted the reader appraised with each matter without subjecting him to that unnecessary circumlocution so common with many writers. He intended to give details of appropriate manures for each crop. Pt. II from p. 10 outlines each manure and its uses. Before that Section III opens with Adam sent forth from the Garden of Eden to till the ground. He thought that the utilisation of some kind or another of sewage must have recurred throughout the primeval ages of mankind and makes a dubious assertion that Noah must have buried the original vegetation, presumably in cultivating crops. He speaks of the value of the Nile floods to the Egyptians and refers to the classical writers, and to the use of prepared chalk and sea water in his own country in the early 18th century, but he wanted farmers, if any of his readers were farmers, to understand that in no trade was duplicity more inherent than in the manufacture of artificial fertilisers. Many worthless preparations were advertised. Section IV of his book was a classification, 1st Vegetable manures, 2nd Animal, 3rd Miscellaneous. He noted many analyses of the suggested fertilisers, each material being dicussed separately. It must have been a useful book.

Crews also wrote *The potato and its cultivation,* also published by Cox, The Field, an 8vo of 51 pp. Crews prefaced the pamphlet by saying "The subject of potato cultivation occupies at the present day a position of great importance in the role of the British Agriculturist and Horticulturist. They successfully compete with their foreign contemporaries . . ." (but) British agriculturists only possess a knowledge of the outer mind of the matter. He gives "thirty-one uses of the potato", some very odd indeed, an example being the proposal that mixed with stucco, the potato would form an improved plaster. Not only does Crews propound the history of the plant but also its chemical construction, and its varieties, English and American. He gives generous and useful instructions for the cultivation of the plant.

One Henry Tanner, sometime Professor at the Royal Agricultural College wrote *The Abbotts Farm*, a small 8vo put out by Macmillan in 1881. In the preface Tanner asked the unanswerable question "Does instruction in science make a man a more successful farmer and a better neighbour?" This book was a collection of sketches which originally appeared in the Preston Guardian. It is not a textbook but a discussion of the serious value of farming both to the farmers and the farmers' wives.

Sometimes it is difficult to decide whether to include a book of foreign origin that was also put out by a London firm. The balance is perphaps in favour of adding something about one such to these pages. The example I have in mind is a work by Lewis Sharpe Ware, *The Sugar Beet: including a history of the beet sugar history in Europe; varieties of the sugar beet, examination, soils, tillage, seeds and sowing . . .*, an illustrated 8vo with a pleasant frontispiece of some 325 pp. issued by H. C. Baird of Philadelphia and S. Low of London.

The preface declares that "It sufficed to say the author found the work a pure labour of love" (what author doesn't?). Ware's introduction admits that "The author had the advantage of advice of Payen and Dumas as well as other eminent chemists when in Paris." During his 14 years' residence in France and Germany he visited the most important sugar beet establishments and stored the practical and theoretical information gathered for further use. Eventually it was suggested the entire subject be written up. There were no authors to consult in the English language, an additional inducement to write the book. A Mr B. Grant had written a small volume in which lands in the Western States were recommended. Professor Crookes' book said little about the actual culture of the beet itself, Ware was not content with this production. He said that an extended treatment could have been written. I do not understand whether he was prepared to undertake this task, but so far as I know he did not do so, or if he did the book was not included in the British Library collection. Ware also wrote a two volume work on Beet Sugar Manufacture but that is outside the scope of my plan of this book.

John Prince Sheldon was not only a prolific writer himself but he was evidently a well known man who could call upon many of his distinguished contemporaries for help in one way or another — in giving him advice and in writing parts of his planned works or even in devising sections of his compilations. One of these which has just this character was *Dairy farming: being the theory, practice and methods of dairying.* This work was a 4to of 570 pp. adorned with a frontispiece, 25 plates and other illustrations. It was published by Cassell in 1880. It has been assessed by a modern critic as "A comprehensive work on all aspects of dairying and milk processing." It discussed breeds of cattle, milk, cheese and butter and remarks that the carriage of milk by rail was a new, recent development. It gives various statistics of production, the animals, the milk they gave and the making of butter and cheese, and he writes of the use of these articles in human diet. The coloured illustrations are very attractive, and an interesting soil map is a notable feature.

Sheldon went on writing and lived into the twentieth century. It is unfortunate that writers no more than anybody else do not live in precise decades. Sheldon was no exception to this circumstance, unfortunate as it may be to a bibliographer. In 1893 he published a book of 154 pp., an illustrated 8vo, one of Bell's Agricultural Series. This was a second edition. I have not seen a first. In the same year he embarked upon the turbulent seas of politics with *The future of British agriculture; how farmers may best be benefitted,* 158 pp. 8vo published by W. H. Allen London. This was a very productive year.

Sheldon also published *British Dairying: a handy volume on the work of the dairy farm,* another illustrated 8vo, published by Crosby Lockwood. The Royal Agricultural Society also has *Practical dairy farming,* another London 8vo, 1893, but this imposing list excepts *Livestock in health and disease; the breeding and management of horses, cattle, sheep, goats, pigs and poultry. With chapters on dairy farming,* edited by J. Prince Sheldon, a 4to of 627 pp. with a frontispiece, plates and other illustrations, published by Cassell, London, 1902. It is an imposing, exhaustive and impressive work but it is a question whether very many working farmers bought it, though buyers must have been sufficient to make the thing a profitable proposition. It would be imposing on any possible readers of this production to attempt to make an exhaustive analysis of this voluminous work, but its quality and character are excellent and it must have been useful if only for reference purposes.

The series *Farming for pleasure and profit* was really by Arthur Rowland. One was *Dairy farming and management of cows,* a Chapman and Hall, 8vo of 210 pp. A second was *Poultry keeping* pp.162, 8vo, by the same publisher. This opens with a discussion of Ancient domesticated poultry. The editor believed the dove sent by Noah from the Ark was a tame one and that in Ancient Egypt, with the increase in poultry keeping the ancient practice of capturing wild birds and fattening them for the table had almost entirely ceased in England, though there was still a stray decoy or so in the Eastern Counties. Quails were sometimes fattened as a peculiar luxury. There is a great deal of miscellaneous information in the book, one example of which is a description of cock fighting amongst the Malays, an odd thing to be supposed of interest to English farmers.

Another volume written by this author but perhaps not of the same interest was *English trees and tree planting,* an 8vo of 434 pp. issued by Smith Elder of London of 1880. Much the same sort of thing was *Arboriculture for amateurs . . . planting and cultivating tress,* an 8vo of 120 pp., copiously illustrated, and containing besides plantations advice about orchards and hedgerow planting. It was published by the Bazaar Office in 1880.

It is perhaps not necessary to include here his pamphlet *Advice to youths about entering a commerical career,* an 80 pp. put out so long before as 1867 by Clark and Co. of London. This I have not read.

One White Stapley wrote *Plain letters to a Tenant Farmer under Ecclesiastical Landlords,* a 16mo issued by a press at Andover. This writer had a mercy upon the church as a Landlord. Ecclesiastics were invariably the worst landlords, having only a life interest in the farms. Here examples are given concluding that much property held by church authorities was left by pious forefathers for specific purposes carried on before the Reformation, not after. He asks "Are these authorities really entitled to some of the land", but this is the sort of controversy that is not purely or indeed partly a description of farming procedure, former or current, and so this writer may be dismissed.

The teaching at the Science Museum, South Kensington, encouraged the production of a number of textbooks whose authors wished to assist in the education there provided for aspiring farmers or perhaps stewards, bailiffs and land agents. One such textbook was written by Hugh Clements. It was *The fields of Great Britain; a textbook of agriculture adapted to the syllabus of the Science and Art Dept., South Kensington for elementary and advanced students,* a small 8vo of 364 pp. published by Crosby Lockwood, London, 1881. The author also wrote a book about *Original Chemistry* and was a lecturer in agriculture etc. at the South London Training College, Blackfriars. An introduction was provided by one H. Kains Jackson. The book is a reprint of material originally published in the periodical *The Farmer.* It opens with a chapter on soils and proceeds to discuss suitable soils for various crops, irrigation, drainage, inplements and machinery, building, manure, and proceeds to discuss all the usual subjects contained in a farmers' textbook on husbandry.

At times Clements becomes philosophical e.g., "The soils that are found in this and other countries has been formed by the disintegration of the rocks composing the earth's crust" and so on. He then sets out the soils suited to the growth of various crops, treats of drainage, implements and machinery, farm buildings, farmyard manure, artificial fertilisers, rotations, seeds, malt, in fact the whole subject including grass, hay, orchards and fruit etc., Cattle, Sheep and pigs, the Dorset breed of the last being black and handsome.

In 1881 John J. Pilley wrote *The 100 elements of scientific agriculture for students and farmers* with an appendix, an 8vo of 272 pp. illustrated, published by Geo. Gill and Sons, London. J. Pilley said in his preface: "As far as possible the work as been written to meet the requirements of the Science and Arts Dept., South Kensington, but occasionally goes beyond the syllabus." Pilley also tries "to arrange the matter . . . that it may tend to awaken and to certain extent to satisfy the curiosity of those engaged in agricultural pursuits" (in fact he wanted to supply both these possible markets). Modestly he acknowledged his debt to such authorities as Johnson, Liebig, Stevens, Way, Scott-Burn, Mechi, Lawes and Gilbert and to the works of Professor Tanner. He signed the Preface as for Charterhouse School of Science and Art, where he lectured.

The various processes employed in farming are described, as are all the other methods which are necessary during the farming year, in unequivocal language and his authority is commendable. To examine the contents of the book in detail would be tedious. They are much the same as those of other textbooks, not only those intended for students but also those intended to instruct or provide a reference to all the varied processes necessary in the mixed farming that was pretty well the general practice at that time, the exceptions being the hill farms when sheep breeding or cattle grazing was the predominent occupation of the moral population.

John Prout of Sawbridgeworth, Herts. was born in Cornwall and was brought up to farming under his father. He emigrated to Canada when he lived from 1832 to 1842, when he returned to England and joined an uncle in a business in the Strand; but town life only held him for a couple of decades. In 1861 he bought Blount's Farm, Sawbridgeworth, Herts., which he cultivated till 1894 — 33 years. The DNB says to him belongs the credit of teaching a practical lesson in scientific farming based on 33 years' successful farming at Blount's. His system was based on his Canadian experience and his study of Sir John Lawes' experimental plots at Rothamsted. He demonstrated that successive crops of cereals could be grown on heavy clay if drained well and deep, and ploughed, being dressed with properly prepared chemical manures. This description of what he did is set out in his book *Profitable clay farming under a first system of tenant right*, the 3rd edition of which, an 8vo of 100 pp. illustrated with plates, came out in 1881. His work on his farm and his writing about it aroused a good deal of controversy and discussion in the agricultural press, for example the Royal Agricultural Society Journal of 1875 contains a "Report on Messrs. Prout and Middleditch's continuous growing." Prout bought Blount's and Sweetdew's Farm in 1861, some 450 acres in a very bad state, which is confirmed by the introduction to Prout's book. He relied on Vocleher's advice (Vocleher was at Rothamsted). He used steam tackle, the then advanced thing to do. He supplies tables of crops grown consistently with his theory of continuous cropping, as the proper and up-to-date method at that time.

Some of the Victorian writers were exceedingly verbose, or would a better word be voluminous, but how could anyone plead guilty writing under (or is it over) the title *Book of Farm Management and Country Life. A Complete Encyclopaedia of rural occupations, With numerous illustrations*, a thick 8vo of some 1370 pp. This was put out by Ward and Lock. It was furnished with a coloured frontispiece showing farm lay out, fields and buildings. It was not dated but was issued at about 1881. Page 1 shows the Anglo-Saxon plough, 142 the Roman plough, 110 the use of the plough in Ancient Greece, instructive but perhaps a little tinged with imagination. After legalitites and so on this author goes on to discuss the farming customs of every country, especially the relation of landlord and tenant in detail as seen in different countries, a discourse that could have been very useful for legal students. These preliminaries being disposed of, he lays down rules upon how to choose and manage a farm, an essay furnished with numerous illustrations of farm implements. Then follow the leading principles of agriculture, where he mentions Tull's theory and Grisenthwaite's theory that the earth was like the stomach of an animal. The soil furnishes nothing to the plant, or so he says, and he suggests a theory of complete soil pulverization as the source of vegetable life; but it would be easy to expand notes on the contents of this work, which closely follows its title. One thing of interest to the historian is a page (132) of illustrations of old husbandry implements. The dairy farm is minutely analysed, and provided with spacious accounts, many illustrations of cows of different breeds, and advice on the selection of cattle for dairy purposes. Sheep follow, then pigs, asses and mules in various chapters.

The horse! — this writer relies upon the Bible for the history of the horse, and proceeds to Cyrus and the Persians. He thought that the horse was originally derived from those portions of Africa nearest to Egypt and adjacent portions of the interior whence he gradually found his way to Arabia, Persia and afterwards Greece. There were five different breeds in England, Cleveland, Lincoln and Suffolk, the Clydesdale in Scotland and the native Irish garrom.

Poultry, turkey, ducks, all were nicely illustrated and carefully discussed. Gardening in all its aspects is carefully treated, sports of all kinds are described, the pursuit of game not being neglected. Trees and their management follow, and finally the dog, its life, training etc. The title of this book said it was a complete cyclopaedia, a claim that it certainly seems to fulfill. "They were giants in those days."

George Bunyard, a Fellow of the Royal Horticulturual Society wrote *Fruit farming for profit: a practical guide on all matters relating to this important industry* . . . an 8vo of 79 pp. put out by Frederick Bunyard of Maidstone and Stanford of London. This ran into four editions between its first appearance in 1881 and 1899. The author states in his preface that he had spent twenty years managing the largest fruit tree nursey in the Kingdom. He was a member of a firm of nurserymen who sold plants and possibly seeds. The book is purely and precisely technical and must have been of great service to its readers.

In 1892 Bunyard published another book, the title of which is *Modern Fruit Culture. Hints for amateurs and others,* being a reprint of the cultural articles from George Bunyard 60's catalogues. It was in two parts, an 8vo published in 1892. It is said that he also produced textbooks on Raspberries, Strawberries, Apples and Pears, Cherries, England's national flower (what was this? the Rose?), Gooseberries and currants for garden and market, but I think the growth of currants in this country doubtful.

The elementary principles of scientific agriculture are the same at one age or another, but the Victorian flame of enthusiasm for education encouraged the production of a plethora of textbooks on one subject or another, amongst which farming played an outstanding part. Many claimants to literacy notice produced textbooks for the use of the humble student. Two of these in partnership were John Charles Buckmaster of the Science and Art Dept. who joined J. J. Willis of Rothamsted in a work with the above title, an 8vo of 199 pp. issued by Simpkin Marshall in 1882.

The preface states "A good part of my early life was spent on a heavy clay farm in the valley of the Chiltern Hills where my ancestors for more than 250 years had been farmers" (presumably Buckmaster's). He had worked for some time with the, then, late Professor Way. Modestly he wrote "I refer to these matters to show that I am not altogether ignorant of the general practice of agriculture." This book is "confined chiefly to the matters laid down by the syllabus of the Science and Art Dept." I may be excused for the belief that it was mainly the work of Buckmaster, the Rothamsted reference being mainly for prestige.

When I was a humble servitor of the then Board of Agriculture, the son of James Long was the Editor of the Board (afterwards Ministry) of Agriculture. This doubtless gave him a certain prestige as the Editor of the Journal and head of the Publications Branch of the Board. Perhaps the notoriety (fame!) of his father helped him, but be that as it may, his father was a most productive writer whose productions were so numerous that I am uncertain and unsatisfied that I have collected even the titles of all his works. This man was Merlin of the Field, though I do not know quite what that implies! At any rate he produced a volume of 372 pp., an 8vo put out by Smith Elder in 1881 under the title *Farming in a small way*. I do not propose to criticise this, or to sum up the value of his other works: that might be called a work of supererogation. His next production was *British Dairy Farming to which is added the chief continental systems,* an illustrated 8vo of no less than 536 pp. adorned with illustrations and published by Chapman and Hall, London 1885. I do not think Long was a man of original or inventive mind, judging perhaps ungraciously from my association with him as Editor of the Board's Journal, but whatever else he may have been he was certainly industrious and prolific.

His next book was put out with the informative title *The Book of the Pig; its selection, breeding, feeding and management,* some 360 pp., adorned with a frontispiece and plates and other illustrations in the text. It was published in 1886. A second edition of this book was published as late as 1929. I do not prepose to discuss its contents which were quite commonplace and equally quite usual.

This man's books were evidently popular and some went into several editions. For example with J. C. Morton he wrote *The dairy farm* some 115 pp., 8vo, adorned with a frontispiece and plates, the first edition of which was published by W. H. Allen of London in 1889, a second edition being called for (presumably) and published in the same year. A 3rd edition came out in 1892.

In 1896 Long collaborated with a John Benson, who is otherwise unknown to fame, in an 150 pp. 8vo entitled *Cheese and cheese making, butter and milk. With special reference to continental fancy cheeses*, which Chapman and Hall brought out in 1896. I think it needs no more than inclusion here.

James Long in co-operation with John Chalmers Morton contributed *The dairy* to Morton's *Handbooks of the farm*, some 148 pp., an illustrated 8vo in 1892. This had previously appeared as one of *Morton's Handbooks* in 1885. Much the same material is to be found in *Modern dairy farming: a manual for all who are engaged or about to embark in the production, manufacture or sale of dairy produce*, an illustrated 8vo of 131 pp. published by the Baxaar Exchange and Mart Office so late as 1916. W. Collins of Glasgow published Long's *Elements of dairy farming*, pp. 208, 8vo in 1894.

This writer was so prolific that it would be beyond the scope of this endeavour to attempt a critique of each of his numerous books, but a note of each title must perhaps be made in an attempt to be inclusive on this bibliography, e.g. *Goats, their varieties and characteristics. How to rear for profit by both milk and flesh*, an 8vo. *Making the most of the land*, pp. 282, 8vo, 1913, *Modern Sheep farming*, illustrated pp. 102, 8vo, 1917. *Poultry for prizes and profit* an illustrated 8vo 127 pp., the 2nd and 3rd editions of which came out.

The story of the farm and other essays is a collection of fugitive pieces that were thought by the author to be worth re-publishing as a collection, an 8vo, London 1898.

Long went to Canada and came home to write a description of what he saw in a volume, *Canadian agriculture, report on a visit in 1893* pp 28., 8vo, 1896. A year before this journey Long wrote *A handbook for farmers and small holders* 227 pp., 8vo, which was published by Sampson Low and Co. The contents of this kind of work can be so easily surmised that it is not necessary to expand upon it here.

I think it extremely unlikely that I have been able to consult and remark upon the total output of this voluminous scribe, but I must confess my human weakness in becoming rather bored in the attempt to do so. Finally I must say that Long was not an original. He followed the trodden paths but he was enormously industrious in a pedestrian way, but then his age was pedestrian.

A hardly less voluminous writer was Sir Arthur Herbert Church, M.A. etc. One of his earliest productions was published in 1882. It was *The laboratory guide, a manual of practical chemistry*. This was a fifth edition and I regret I have not been able to place the earlier ones. It was a 266 pp. issued by J. van Voorst in 1882. My own comment on this book in my card index is that it needs no comment, so I shall not make any. This was succeeded by a 12 pp. pamphlet on *Sulphate of Ammonia as a manure*, an 8vo printed by G. Slater of Sheffield in 1886. The Bath and West Society's catalogue states that A. H. Church was the revisor of S. W. Johnson, *How crops grow etc.* This was a *treatise on the chemical composition, structure and life of the plant for agricultural students*, an illustrated 8vo of 399 pp., published first by Macmillan in 1869, wherein the author is given as Samuel W. Johnson, it being revised and with numerous additions by Arthur Herbert Church, Professor of Chemistry at the Royal Agricultural College, and William Thiselton Dyer, Professor of Natural History at the same place.

The introduction is in my view a glimpse of the obvious, but this was apparently not so to the writer. He states pontifically that "The objects of agriculture are the production of certain plants and certain animals which are employed to feed and clothe the human race". The first aim in all cases is the production of plants (an observation I had not seen before nor heard) but true for, as the preacher said, all flesh is grass. Man has to cultivate to live, which Church condemns as "In this defect or rather neglect of nature (I suppose he means to supply sufficient natural provisions?) agriculture has its origin . . ." Every successful farmer", he wrote, "is to some extent a scientific man. The farmer without his reasons, his theory, his science can have no plant . . . The more he knows the more he can do".

"Agriculture was practised hundreds and thousands of years ago. Nearly all the essential plants of modern cultivation were known to the Romans before the Christian era. The writings of the Chinese show that their wonderful skill and knowledge were applied to agriculture at a vastly earlier date . . ." Church speaks of Dundonald as the originator of agricultural chemistry, and proceeds to cite Cato and Columella. He wrote rather patronizingly, and knew something of manuring.

He notices the Agricultural Chemistry Association of Scotland of 1843, which after five years of existence was merged in the Highland and Agricultural Society, an Association that was headed then by James F. W. Johnston followed by Dr Anderson.

The book is in three parts. Div. 1, Chemical composition of the plant, II, The structure of the plant and offices of its organs; III. Life of the plant. It has as appendix of some 20 tables plus 64 illustrations. A notable work which is written in a most attractive manner, though of course its subject matter is much the same as that to be found in the numerous agricultural chemistry textbooks of the period.

Few agricultural pundits of the latter part of the 19th century have been distinguished by an entry in the DNB but John Elton Taylor achieved this honour. By the same token it must be admitted that the author's agricultural writings seem to have been limited to a couple of brief pamphlets, but then he was acclaimed as a popular science writer. Taylor was a man of consideratable enterprise as so many of our 19th century ancestors were. Although he was brought up to work in a humble capacity as a storekeeper on the L. and N. W. Railway, he was so praiseworthy as to arouse the patronage and encouragement of one Mr Ramsbottom. He knew a sub-editor of the Norwich Mercury under Richard Bacon. He is said to have become a popular lecturer on Science and to have established Norwich Girls School in 1864 but what is of immediate interest here are two pamphlets, the one *How to profitably improve our soils* a 16 pp. 8vo put out by E. Packard, Ipswich 1882, and *How plants grow. With a preface by Augustus Vocleker,* a 29 pp. illustrated 8vo put out by Packard in the following years. The titles of these pamphlets are perhaps sufficient indication of their contents: therefore I shall say no more of this man's contribution to the literature of farming.

Rather infrequently it is difficult to decide whether to include a book or not. George S. Heatley poses a problem. His book *The stock owners' guide, A handy medical treatise for every man who owns an ox or cow,* an 8vo of 172 pp., is really a veterinary treatise, and should I think need no more than a cursory mention here; so I will be silent about him.

Another book, a copy of which is filed at Southampton, we could not find in the British Library. It is *Practical Farming,* an 120 pp. 8vo printed at Horsham in 1882, the author being one A. F. Parbury, but this we have not seen.

Several of the unidentified contributors to the *Field* newspaper who produced a 44 pp. 8vo supplied with some illustrations in 1882 were apparently guided by one who called himself Agricola. This work was entitled, *Harvesting crops independently of the weather: practical notes on the Neilson system of harvesting.* These inspired prophets remarked that "Enormous losses have been suffered in all times by the bad harvesting of grain and fodder crops . . ." Happily (with astonishing modesty) the discovery has at last been made that hay and corn may be harvested in continuous raining weather as perfectly as when the skies are clear and the sun shines brightly, in the Neilson system.

"According to the method which has taken the world so much by surprise (whatever small proportion of the world ever heard of it?) the grass makes itself into hay by that law of chemical action which causes vegetables and other perishable substances if put together damp or green to ferment and generate heat — the temperature however being regulated by drawing off from the rick all steam and excess heat above a certain range ascertained by means of a thermometer. Thus grass does not need to be sun dried but after being slightly withered will make itself into hay as good as can be secured by the old way of hay making under favourable conditions and consequently of far better quality than what is usually made in bad weather." This was certainly a generous promise but seems a bit unlikely to be fulfilled in the physical world, pleasant as that would have been.

One T. H. Baker compiled *Records of the Seasons, prices of agricultural produce and phenomena observed in the British Isles.* This is not precisely a farming book or even a book about farming but it is worthy of mention here if only in admiration of the industry of the author, and indeed because the seasons are a factor of the utmost importance to the farmer, to give a glimpse of the obvious. Baker was a well-read man if he may be judged by the quotations he makes in his preface. The book was an 8vo of some 360 pp. published by Simpkin Marshall, London, and B. W. Coates of Warminster in 1883. The citations to which I refer are of Gilbert White, Hume and Holinshed with Lowe's *Natural Phenomena,* a book with which I am not conversant. Baker makes some very odd remarks but is particularly interesting in his attention to and recording of "Great Frosts". For some reason, not immediately apparent to me, the great outbreak of plague in 430 A.D. is described, if that is the proper word. More important to the historian are the modest details of grain prices at various times.

In 1883 still one George Heatley Smith induced (authors always persuade publishers!) W. H. Allen of London to put out his book *Sheep farming*, some 254 pp. accurately illustrated with plates, an 8vo. The author's preface informs the reader that "This work enters minutely into the history, progress and development of the various breeds of sheep. It specifies in plain, explicit language the many ailments to which sheep are liable and recommends the remedies that are most efficacious and suitable for their cure".

"The work will lighten the difficulty that is experienced by the owners and attendants of this important animal not only with respect to disease but also to the best means of prevention which is infallibly better than cure".

But this book seems on consideration to be rather a retiring veterinary treatise rather than a flock master's procedure guide, and should perhaps therefore be left at that because this study pretends to be devoted to farming treatises. He wrote other books about the ailments of horses, dogs and stock.

The name of Primrose McConnell is so well known and his note book in a modern edition is still a handbook for farmers, so I shall confine myself here to a short bibliographical list of his contributions to knowledge, or if not that, his useful dairy references for farmers. In 1883 Crosby Lockwood published his *Notebook of agricultural facts and figures for farmers and farm students,* an oblong small 8vo of some 538 pp. He wrote a few other things, a 27 pp. 8vo on *The agricultural depression*, not of course a textbook but a political pamphlet issued by McCorquodale, London in 1887. These people also published a 23 pp. pamphlet, 8vo, on *Agricultural Education* in 1890. McConnell also wrote one of Morton's *Handbooks of the Farm, The elements of farming* (q.v.), and as late as 1902 *The elements of agricultural geology, a scientific aid to practical farming,* an 8vo.

The series *Handbooks of the Farm* was not invariably put out by the same publisher (printer?). For example Maxwell Tylden Masters wrote No. 5 (2) of the series with the title *Life on the farm, Plant Life,* an 8vo of 142 pp. printed by Bradbury Agnew of London in 1883, a second edition of which, similar in pagination and size, appeared in 1885. Masters was apparently mainly an horticulturist and possibly a botantist. The general editor was, as of other publications in the series, J. C. Morton. I must admit that I have not read this book (mea culpa!) and so cannot make any appropriate description of its contents or character.

J. E. Thorold Rogers is much better known (if he is not now forgotten!) as a historian than as a writer on farming subjects, but he did write on ensilage then and slighty takes a "burning topic" amongst the rural population. His 103 pp. illustrated 8vo of 1883 was put out in London. It was *Ensilage in America. Its prospects in English agriculture.* It was not of course the first work to deal with silos! I have seen in a secondhand bookseller's catalogue, but no closer than that, *Ensilage, instructions how to build silos,* a title only to me. The citation runs, Ensilage having been practised is France was the subject of a treatise published in 1877 by M. Goffert, a work that was translated and published in America by Mr Brown (otherwise unidentified) in 1879. The practice was taken up euthusiastically and Thorold Rogers recommended it or so says Ernle on p. 387: "Ensilage was warmly advocated in 1888."

Thorold Rogers does not depend for his reputal upon this somewhat minor work but upon his two massive productions, *A history of prices and wages in England from the year after the Oxford Parliament (1259) to the commencement of the Continental War 1793,* seven volumes in eight, 8vo, Oxford 1866-1902, and *Six centuries of work and wages, the history of English labour* 8vo, London 1894. In that year too appeared *The industrial and commerical history of England* Edited by A. G. L. Rogers, 2nd ed 8vo 1894. This is not the place to comment upon these works but is one when they must be mentioned.

It seems to me to be a bit invidious and rather in the way of denigration to call a writer a compiler, but I suppose that is in effect what we all are. Certainly it is true of Frederick Woodland Toms who produced an illustrated 8vo of 219 pp. in 1883 entitled *Silos for preserving British fodder crops stored in a green state. Notes on the ensilage of grasses, clovers, vetches, etc. Compiled from various sources by the sub-editor of the Field;* this carried the imprint of H. Cox, the Field Office, London 1883. A second edition came out in 1884 and a third in 1885. Parallel to the second there appeared *Short notes on silo experiments and practice. Extracted from Silos for preserving British fodder crops,* a 40 pp. pamphlet, 8vo in size also issued from the Field Office. It is perhaps unnecessary here to expand upon the methods of making silage, which is after all the same today as it was yesterday and merely requires a container in which to store the grass or a carefully composed pile or heap of this material.

Thomas John Elliot was the Professor of Estate Management at the Royal Agricultural Collge at Cirencester and a resident agent at the Southwick Park Estate, Hants. He wrote *The Land Question; its examination and solution from an agricultural point of view as illustrated by 23 years experience on the Wilton House Home Farms, Salisbury, Wilts by Lord Herbert and Lady Herbert of Lea. Analysed by T. J. Elliot*, an 152 pp. 8vo issued by Cassell in 1884. A plan of the farm forms the frontispiece. It is an exhaustive and informative work, including an analysis of 23 years of accounts covering the years 1850-1873, although the Professor himself says that the book was in a great measure the subject matter of lectures given at Cirencester. The accounts show that farming was profitable from about 1850 to 1873. The farm described was a mixed farm carrying sheep, cattle, swine and poultry. He made an experiment in fattening cattle. The book is very detailed and supplies a miscellary of statistics such as crop yields, horse keep, labour and, rather esoteric, the damage done by game to crops. The accounts are set out in great detail but much of the British Library copy is still uncut and awaits the patient and detailed study of some as yet unknown but aspiring student. Need I say more!

The History of the Clydesdale Horse of 1884 is mentioned in the Catalogue of the Highland and Agricultural Society Library but there is no more than the title in this entry and I have not seen this book.

Another puzzle is a large folio, 30 sheets of mounted grasses, entitled *Specimens of British grasses* put out from Dublin in 1884 which does not seem to be in the British Library Catalogue but there is a book in that collection by one Edward Hackel entitled *The True Grasses,* a translation of *Die naturlichen Pflanzen*, G. F. Lamson Scribrer and Effie A. Southworth, an illustrated 8vo of 228 pp. published by A. Constable, Westminster, 1896. Long before this, in 1846 (see *Old English Farming Books. Vol IV, 1840-1860*, 1984, pp. 40-41) Frederick Hanham had edited *Natural illustrations of the British Grasses* put out by Binns and Goodwin of Bath, in 1846, garnished with specimens mounted on the plates but some of them have vanished.

A simpler and more straightforward author was George Nevile, who was a man with a simple sense of humour. He wrote 234 pp. illustrated by plates and sketches in the text, an 8vo published by Longman, London 1884. It had the simple but explicit title *Farms and Farming,* "In treating of a subject so thoroughly threshed out, so to speak, as farming it is difficult to lay claim to originality." He was justly proud of his chosen employment and of his country and did not hesitate to say so. "The agricultural industry, the greatest industry of one of the greatest nations of the work" needed, he thought, "some stimulus to preserve it from decay", and this he proceeded to supply.

Part I of the book deals with the capital required, the plants cultivated in ordinary husbandry, manuring, rotation, and this part also deals with the food of plants, manures natural and chemical, laying down to grass, high farming and low farming, varieties of soil and trying experiments. Part II is on breeding animals, cattle, sheep, pigs and farm horses. He deals also with Cross breeding.

Part III covers Farm buildings, with some plans, Dairy Farm accounts, Implements commonly used, ensilage etc.

Part IV discusses the laws relating to capital and labour. Part V, Farm horses, in which he refers to Youatt's *Book of the Horse* (see *The Old English Farming Books. Vol. III. 1793-1839,* London 1983, pp. 168, 169).

In general the book may be said to deal with what may be called the common methods, rotation of corn and grass and so on. He exhorts his readers to take care when making experiments, and to consider all the factors involved on taking advice.

Charles W. Stubbs did not content himself with giving written advice to the farmer or to the labourer but took the practical measure of setting up some allotments to be rented by the local helots. This experiment is described in *The Land and the Labourer. Experiment in cottage farming and co-operative agriculture,* an 8vo of 186 pp. and some tables, published by Swan Sonnenschein & Co. in 1884. He described himself as a Country Parson upon whom two conclusions were forced by the conditions in which so many of his more humble parishioners lived after he had spent twelve years in his parish. The first was the revision of the English land system to raise the social and economic condition of the English rural labourers. This is impossible unless (a) either there is an increase in the proportion of small agricultural landholdings in England or (b) the adoption of methods, probably co-operative, which shall make it economically advisable to increase the amount of English labour on English land.

At the end of 1873 Stubbs divided a portion of his glebe some 22 acres into 1/2 acre allotments, annual rent 66 shillings an acre, retaining one acre himself. All of them were worked on exactly the same system of husbandry, accurate accounts of outgoings and incomings being kept of his one acre. The results in the then previous six years of agricultural depression was that the net profit on the acre, after allowing for rent, taxes, seeds, labour and manure was £3.8.0. This was certainly excellent and I am sure that many a contemporary farmer would have been pleased if not indeed gratified if he could have made the same profit on each acre of his farm, which was to my mind a bit unlikely.

The Centenary of the Highland and Agricultural Society of Scotland was in 1884 and in that year a *Historical Sketch of the Agricultural Progress of Scotland* was published in Edinburgh. This is not a farming book in the sense which I use the term but is very worthy of being mentioned in this place.

I am afraid I shall have to be almost equally brief about a book which we could not find in the British Library but which adorns the shelves of the Perkins Library at Southampton. It is *Among the Clods or phases of farm life. As seen by a Town Mouse,* an 8vo of 328 pp., printed by Timley, London 1884. It sounds very attractive.

Henry Woods was agent to Lord Walsingham, a situation which gave him every opportunity of observing the practice of farming in his neighbourhood. The result was the production of a brief work on *Ensilage: its influence on British agriculture; also Southdown Sheep their history breeding and management. Two lectures delivered at South Kensington Museum enlarged and revised* pp. 53-26, 8vo, Hamilton Adams, London. This was published under the auspices of the Institute of Agriculture. Before this Woods had given *Lectures on the breeding and management of sheep delivered before the Wayland Agricultural Association* which was published as a London 8vo in 1864. This, he said, he had been pressed to print and he assured his readers, if any, it was only his own experience that he now offered to the public. The principal points dealt with can be set out as the author did himself i.e. (1) The ewes to breed from (2) Rams, how to use (3) Treatment during pregnancy (4) Abortion, its cause and effects (5) Management of lambs (6) Hoggets from July to Michaelmas (7) Hoggets on turnips (8) Whether more profit to sell in or out of wool. It hardly seems necessary to give any more details of this brief work but is sufficient if no more to quote Bell's *Weekly Messenger,* which pronounced that "All flockmasters ought to be thankful to Mr Woods" and printed a letter from one John Hammond, and Woods' modest answer.

The book was translated into German as *Uber die Züchtung des Fleischschafe. Ein Vortrag* . . . by N. M. Witt, Glogan, in 1865 in 8vo format.

Woods also produced *A lecture on the diseases of sheep,* revised and enlarged, an 8vo of 1873.

Robert Warrington the younger was a distinguished man in his own right as well a having been Professor of Agriculture at Oxford. He contributed to Morton's *Handbooks of the Farm* series *The chemistry of the farm,* the first edition of which seems to have been an 8vo (I have not seen it!) London 1881; a 2nd edition came out in 1882 and a 3rd, pp. 128 8vo, Bradbury Agnew. This went into many editions, a 9th, 8vo, London 1894, and reprinted in 1895; a 15th edition (no less), an 8vo, London 1902. A pamphlet of 24 pp. 8vo which was entitled *Brief notes on the physical and chemical properties of soils* . . . was put out in 1894. This subject was expanded by Warrington in *Lectures on some of the physical properties of soil,* a 231 pp. 8vo adorned with a frontispiece which is a portrait of Joannes Sibthorpe and other illustrations, published by the Clarendon Press, Oxford in 1900. He declared in his introduction that the physical properties of soil and their bearing on fertility were a much neglected topic. Schubler was the only early researcher. This book is highly technical and sets out the science of that day, an exhaustive if rather pretentious work.

In 1894 Cassells published an 130 pp. 8vo entitled *Sulphate of ammonia; its characteristics and practical value as a manure,* which need not perhaps be discussed in detail, the title being a sufficient indication of the contents of the book, which went into many editions, the 9th in 1894, the 15th in 1902. His *Book of Rothamsted Experiments* came out in 1905 but this need not be expanded upon here, or so I think, though it is always difficult to decide how to deal with the productions of an author whose works were dated out of my chosen limits

Another writer's work presents itself, and it was indeed voluminous, with the problem of the time of issue — some of it appearing well into the twentieth century. His name was Sir R. P.

Wright and although he made a major contribution in the number of his publications, I find it quite impossible to assess his output at its true worth. Doubtless it was useful at the time. His major contribution seems to have been (note how uncertain I am!) *Principles of Agriculture. A specific subject of instruction in public elementary schools*, an illustrated 8vo of 144 pp., one of Blackies' *Elementary Textbooks,* which is dated 1884. A revised edition, another illustrated 8vo (edited by R. P. Wright) of some 206 pp. came out in 1891. In this edition the introduction claims that the book had been carefully revised, changes made in the text of most pages, paragraphs rewritten and that Part IV was entirely new and contained additional matter that seemed more fitted to make the textbook more helpful to the students. It is exactly what the author described. A very comprehensive textbook for its day, it gives a concise, easily read description of plants, their needs and way of growth. Soil and its cultivation is followed by questions set by the Science and Art Dept. at then recent examinations. Throughout this work, which was aimed at readers who were elementary school students, emphasis seems to have been based on making the contents plain and easy to the student, a very necessary plan as agriculture is a vast and complicated subject. Each chapter is followed by questions intended to test the degree to which the student had absorbed it.

Of course, as I have said, neither authors' lives nor the publication of their writings conform to the decades of time or even the centuries, and Sir R. P. Wright was no exception. He went on working well into the twentieth century but I do not propose to follow him that far except in a little detail. He wrote *Women's place in rural economy,* 1913, translated Fleishman's *Book of the Dairy,* an 8vo of 1896, with the collaboration of C. M. Ackhurst. An odd thing was *The influence of phosphates on farm crops. As illustrated by the growth of plants in pots in the Glasgow Exhibition of 1901*, a pamphlet of 24 pp., 8vo, printed by John Lewis and Co. which was put out again in the following year. *The standard encyclopaedia of modern agriculture and rural economy by the most distinguished authorities . . . under the editorship of Professor Sir R. P. Wright* came out in 12 volumes 1908-11, but needs no comment here.

The Rev William Holt Beevor wrote four books, an admirable performance though perhaps not so gargantuan as the output of some of his predecessors and successors. The first of these, which is really only a handbook, is *An alphabetical arrangement of the leading shorthorn tribes with notes for the use of breeders,* the second edition, a 304 pp. 8vo, was issued by J. Thornton of London in 1885. This opens with very much the usual remarks on the origin of the breed which he said was imported by the religious houses of the Middle Ages. The imports were a conglomeration of the best of all breeds brought in by the Benedictine or rather Cistercian monks. Beevor cites a long list of his authorities which I do not think it necessary to repeat here.

Some years before this Beevor had published through Bradbury Evans of London an 8vo of 313 pp., supplied with a frontispiece, over the title *The daily life of our farm*. This is in the form of a calendar of weather and work done from October 1865 to August 1870. The land of the farm, meadow and arable was of fair average quality, some very good, some only middling. The stock of all sorts had been carefully selected. This book is very personal and contains anecdotes rather than regular instruction but is an excellent record of experiences which were most obviously enjoyed.

Beevor's third book was *Notes on fields and cattle from the dairy of an amateur farmer. To which is appended a prize essay on time of entry on farms* . . . an 8vo. of 275 pp., issued by Chapman and Hall in 1862. In the preface to this book Beevor confesses, or should I say asserts, that "the following pages are not a transcript of occasional notes made during a few months' study requisite for the profitable conduct of a small farm." Farming at present is assuredly at a premium (an unusual omission for a farmer), in favour from the Queen, who inspects her animals, to the Yorkshire manufacturer of gigantic gains and the agreeable authors of that charming book *Our farm of four acres*. Agriculture was, Beevor declared, the very core of a nation's well-being. The contents of this book are exemplary and exhaustive. In the selection of cows for the dairy all sorts and breeds are discussed, as well as horses. He deals efficiently with all branches and processes of farming both pastoral and arable. The fourth effort of this industrious farmer and writer was *Successful farming, its essentials . . . from the dairy of an amateur farmer* issued by Bradbury Evans London 1870, an illustrated 8vo of 223 pp. This was in the main a reprint of *Odds and Ends* an appendix attached to *Notes on Fields and Cattle* in its first edition, a perfect *vade mecum* for a farmer. Parts had been rewritten and there is additional matter on new subjects (that is new in this work!). This book contains the normal subjects of any good farming textbook then or now with some comments on finances plus a calendar of operation which would be tedious to recapitulate once again although a historian might analyse his comments in relation to the accepted processes of his day.

The Reverend George Brooks wrote what was rather a political exegesis than an agricultural textbook. His excuse for writing the book (if any author needs an excuse) was that he had been induced to take up the cause of the farmers owing to the unjust, inhuman and brutal treatment which many of them had received at the hands of their landlords. He told his readers "You need not be reminded that a severe agricultural crisis exists throughout the whole of Great Britain especially in some parts of the country and more especially perhaps in East Lothian which has for many years had the reputation of being the finest and most highly cultivated country in the Kingdom". He provides many extracts from the contemporary press. The book is, however, political not agricultural. It contains nothing about how to farm. The title of this profound work is *The distribution of Scottish agriculture. a statement of facts respecting the present position of agriculture in Scotland especially in East Lothian with some particulars of the case of Mr James M.Russell . . .* an 8vo of 104 pp. put out by Woodward Fawcett of London in 1885.

There is always a plentitude of people who want to put the world right, and possibly some of them actually do something if only demonstrating how not to do some particular job. One of these philanthropists was Mrs Katharine Burton who not only preached a sort of gospel but practised her own precepts which are set out in *My home farm,* an 8vo of 128 pp. put out by Longmans in 1883. This lady was the wife of one John Hill Burton. She was born an Innes. Mrs Burton's book begins as a letter to M — either a real or imaginary gambit for her book on personal experience. M is thinking of taking a little farm which can be had "for half the rent you pay for a mere town dwelling."

She advised her readers to be as systematic as possible "Pull down nothing till you see your way clear to erecting something better", which is a most admirable sentiment, a maxim to be followed in everything. Hen farms and bee farms are attractive, little or no capital being needed. Their success depends on the farmers' own capacity for profit. A small farm as I would propose does not amount to hundreds of pounds annually. Payments are made in ready money so the ability to lay out £50 annually or at any time is necessary. Thrift is the mainspring of a successful farm but not a narrow of selfish economy. Mrs Burton does not advocate large farms. "Their day seems gone by. They no longer pay. Farming would never be a business for a man possessing above £25,000 of capital."

She makes the obvious remark that good relations between mistress and servant are necessary. The ideal farm should, she remarked, perhaps be cultivated by apprentices but she herself employed a man who could milk, make butter and cheese, cure ham, sew and knit, cut out clothes tolerably well, mow, ride and drive; hatching and rearing chicks was his main talent. "Hens shared his bedroom and over his bed". She gives details which would have been very useful to those new to farming! Finally "Experience and practice each are necessary" but "People with this are not always ready to advise, hence this book in the hope of supplying this defect". This is a very engaging book and its quality has perhaps led me to expand upon it rather largely.

I shall expound much less on William Townsend Carrington who wrote one of Morton's *Handbooks of the Farm*. It was *Livestock of the Farm*, an 8vo of 1882, a third edition of which was published in 1885 (I have not seen a second edition!). It is said that Carrington died while the book was in progress of preparation. The Bath and West Society holds the 1882 edition, a London 8vo of 156 pp. It was the second volume of a series of five others on *Chemistry and crops, Soil and its tillage* and *Equipment of the farm and the estate*. Need any more be said of such purportedly popular pulications even while they were somewhat recondite. The contents must be quite self apparent and need no comment or commendation from me or am I just being lazy? I fear this last!

To turn to a more individual work! Joseph Darby produced a 48 pp. 8vo pamphlet with the expansive title, *Pastures old and new. A plea for the improvement of old for better systems of grassing down and . . . alternative husbandry grass layers*. He said that grassland was often a neglected part of the farm, usually often wet in low sites; for example the Bridgewater Level was infested with buttercups (a beauty for the visiting townee!). Marshy grass too was a necessity for the fluke though some was good e.g. at Romford, on the banks of the Avon, in Hants. In these places the folding was good. In Sussex Sir Curtis Lampson had a lairage for sheep fed on roots and trough foods following the example of the Lawes experiments.

"Scientific men" said Darby "are perfectly agreed that pastures have a great advantage over arable land". They took nitrogen from the atmosphere, he wrote. In Cheshire fertilisers and bones doubled or trebled the produce of worn out dairy pastures. He then went on to discuss forming new pastures, the mixtures of seed to be used and the cost and in Chapter III Faune de Laune's reformed system of seeding, drought resisting plants and the temporary pastures of alternative husbandry i.e. two years and grassing down for a temporary period of from three to six or seven years and for a permanent pasture. This selective work was published by Horace Cox of the Field Office.

A pamphlet that does not appear to be in the British Library was written by one George Fry and a copy is filed at Southampton. It is *The Theory and Practice of sweet ensilage*. It is a 60 pp. 8vo published by the Agricultural Press in 1885. I can say no more about it as I have not seen it.

Thomas Garnett whose *Annual of Philosophy*... is noticed on p. 44 of vol. 3 of this series wrote *Essays in Natural History and Agriculture*, some 241 pp. 8vo, published by the Chiswick Press, London, in 1883. This is largely an exegesis upon his activities which were many and varied. The DNB reports him as being possessed of an enquiring and speculative intellect and an unwearied observer and experimenter in agriculture and natural history. Perhaps it is somewhat exaggerated to note remarking that his papers on natural history and kindred subjects were comparable to Gilbert White.

Occasionally, if not frequently, a bibliographer encounters an insurmountable obstruction to his devoted studies. This is especially true in relation to books which do not appears in the British Library Catalogue. One such is that by J. B. Harris entitled *The cheese and butter makers' handbook: a practical treatise on the arts of cheese and butter making*, a 207 pp. illustrated 8vo put out by Dunn and Wright of Glasgow in 1885.

Another book in the British Library, amongst others of course, was destroyed by enemy bombing in the Second World War. This was *Hops and Hop Pickers*, the People's Library, SPCK, recorded as a 191 pp. 8vo with a frontispiece and other illustrations but obviously not now to be seen. And we could not find in the British Library William Charles Taylor's *An agricultural notebook to assist candidates in preparing for the Science and Art and other examinations in agriculture*, an 8vo of 106 pp. published by Longmans of London in 1885.

I am a little uncertain about herdbooks but I will mention the *English Guernsey Cattle Society's Herd Book*, Vol. I of which came out in 1885 and has probably done so each year ever since. It needs no explanation or discussion for it is what it claims to be.

One of the *Handbooks of the Farm* series edited by J. C. Morton was *The crops of the farm*, 148 pp. 8vo, 1886, garnished with tables. It was a joint production, as the units of such a series nearly always are. This is a technical work written by experts by whom permanent pasture was regarded as the most important part of then current farming. The book deals with each crop in succession, Chapter VIII says, an observation of some veracity. "Weeds of all kinds are robbers." It would be impossible to detail the contents of each unit of this series: many indeed are very similar to one another, even to the use of the same words.

Long before he had made his contribution to this series, Bowich had written a Prize Essay on *The management of a Home Farm* to the Royal Agricultural Society in 1869. This was published afterwards by Tweedie of London and Smith and Son of Dublin as an 8vo of 36 pp. with a frontispiece. This writer was not modest about his literary efforts. He said "This treatise is respectfully issued in the interests of my Business Agency" but he was not ashamed of being obvious. "Home Farms" he wrote "could be good or bad example". He wrote descriptively on notable places like Woburn and Holkham. He completed his work with some details of good buildings and stables. Some years before this he had written an RASE Prize Essay on the rearing of calves which was first printed in the Journal of the Society in 1861.

One Sir George T. Brown of the Agricultural Dept. of the Privy Council wrote one of the *Handbooks of the Farm Series* edited by J. Chalmers Morton. It was *Life on the Farm. Animal Life*, an 8vo of 141 pp., printed in 1886 but this is, I think, rather a veterinary than an agricultural book and so need not be discussed in detail here.

Another of the *Handbooks of the Farm* series was written by William Burness. It was *The equipment of the farm*, an 8vo of 142 pp., the first edition of which came out in 1884. This was published by Bradbury Agnew of London but perhaps this hardly needs stating. A second edition came out in 1886. In the preface the various implements necessary are described as well as the requirements of a farm in horse power, labour and livestock and illustrated by a number of examples. The capital required to finance different sizes of farms is stated. Roads, fences and water supply are detailed as well as the size and kind of homestead suited to different areas of land. A rather odd inclusion is an estimate of the landlords' capital. The book is comprehensive and discusses the then modern developments of equipment and livestock. Some actual farms are used as instructive illustrations, e.g., North Charford Farm, Downton, Sailsbury, a Midland dairy farm, small hill country farms, a Country Farm of 610 acres, a North Lincolnshire Farm, a Gloucester Farm and a Worcestershire arable farm. Roads, fences, drains and drainage form interesting sections.

Burness was apparently a professional author. He had previously produced *The farmers' harvest companion*, which I have not seen. The British Library Catalogue states that a new edition of this was printed in 1870 so I am little dubious about its originality. It may have been a periodical production, annual perhaps? He wrote other books which are not about farming or rural life so are not included here as they deal with subjects outside the scope of this study.

Mordecai Cubitt Cooke proposes still another amongst the many puzzles that offer themselves for solution to the ambitious bibliographer. He wrote a book *Rust, smut, mildew and mould. An introduction to the study of microscopic fungi*, the second edition of which was issued by W. H. Allen of London in 1886 with illustrations by J. E. Sowerby, 262 pp. of 8vo of text, but Southampton lists a 2nd edition of 1870 which I have not seen. There has also been set out by the bookseller, Campbell, of Aberdeen a 5th edition of 269 figures, coloured on 16 plates of which 11 plates were drawn by Sowerby. The Highland and Agricultural Library also sets out *Fungi, their nature, influence and uses*, an illustrated work published in London in 1886. At any rate Cooke in his preface expressed his wish for people to use microscopes to study some of the minute mysteries of nature. He thought it difficult to give a logical definition of what constitutes a fungus. Taking a stroll from London down to New Cross he saw diseased plants on the railway cutting. The coloured illustrations included in the book form a valuable indication of the spores and the plants affected, he having collected, examined and preserved microscopic fungi. The book in my view is a guide to the amateur botanist and has no relation to farming as such, though it must have been of immediate interest to any farmer who happened to come across it.

Hereford and Worcestershire were always, or at least for many centuries, famous for their fruit productions, apples and pears, perhaps from the wilds of Pre- conquest or even prehistoric times but that is pure (if any such can be!) speculation. This specialised line of production is recorded by one Robert Hogg who wrote a 247 pp. 8vo, *The apple and pear as vintage fruits garnished with poetic quotations* which was published by Jakeman and Carver of Hereford in 1886. In his preface Hogg wrote "A century has nearly elapsed since any systematic British work has been published on the apple and pear." He then refers to Marshall's *Rural Economy*, 1789 and Thomas Andrew Knights' *Culture of the Apple and Pear* of 1797 as well as the *Pomona Herefordensis* of 1811. He also refers to a number of continental societies and to the immediately local Woolhope Naturalists' Field Club, then since nine years engaged in obtaining information. He had visited Rouen on behalf of the Club. In that place there had been a Great Exhibtion of Apples and Pears in 1884. The results of all his enquiries, or so he said, were included in his *Herefordshire Pomona* "published at considerable expense" and "containing original figures and descriptions of the most esteemed kinds of apples and pears". This was some two volumes folio published at Hereford between 1876 and 1885. The book was adorned with coloured plates. A large amount of original information is distributed throughout its pages from beginning to end, which its size demonstrates, and makes it very difficult to synopsize but it may be remarked that it includes a list of Perry Pears and Cider Apples from all the five western counties,

James Macdonald (1852–1913) was the Secretary to the Highland and Agricultural Society of Scotland. He wrote the *History of Hereford Cattle,* an 8vo of 380 pp. with plates printed by Vinton, London, 1886, a revised edition of which, edited by James Sinclair, came out in 1909.

Macdonald also wrote *a History of polled Aberdeen and Angus Cattle*, giving an account of the origin, improvement and characteristics of the breed, an 8vo of 459 pp. illustrated with plates which Blackwood of Edinburgh and London put out in 1882. Again James Sinclair edited the *History of Shorthorn Cattle*, an 8vo with a frontispiece and plates of no less than 895 pp. issued by Vinton, London in 1907. Before this he had written the *History of the Devon breed of cattle*, another Vinton 8vo of 392 pp. with plates which Vinton published in 1895 for the Devon Cattle Breeders' Society. James Sinclair also revised a new edition of *Sheep: domestic breeds and their treatment*, with a veterinary section revised by A. H. Archer, one of the Popular Livestock Series issued by Vinton, London 1896. It was an 8vo of 144 pp. with a frontispiece and plates. Based on the work of W. C. Martin, *Sheep our domestic breeds and their treatment,* 1896, it was however so thoroughly overhauled that very little of the writing of the original author remained, or so the preface says. An exhaustive discussion seems unnecessary once they are put on record with sufficient physical notes. Their contents are more or less obvious.

Thomas Potter was Clerk of Works to Lord Ashburton. He appreciated his position in that employment and was sufficiently proud of himself, by no means an unusual phase of character. He thought well enough of his job as to set himself out to form a society of those people who were engaged in the same sort of work. He wrote an unusually lengthy treatise, for its subject, on *The construction of Silos and the compression of green crops for silage,* some 187 pp., an 8vo provided with graphic illustrations issued by Batsford, London in 1886. Over twenty years later (authors will not confine their work to decades!) in 1909 he wrote a London 8vo on *Buildings for small holdings, materials, cost and methods of construction,* a useful enough book but not precisely about farming though concerned with one of its necessary adjustments. Much later he did himself the honour of writing an *Autobiography,* which was published at Winchester in 1924 with an introduction by W. H. Chesson, who is otherwise unknown to fame. One thing is certain. Potter was a man of decisive and emphatic character and his contribution to the knowledge of his day was by no means insignificant. No more need be said about him and his work in this compilation.

Whether or not to include a particular work in this bibliography is a question sometimes puzzling to answer. It is no doubt better to include rather than to omit, so this I shall do with the small pamphlet of 30 pp. 8vo printed by Harrison of London in 1886. Its title is *On the origin of agriculture* and is a reprint from the Journal of the Anthropological Institute, November 1886. This is historical rather than agricultural, as the introduction makes clear when it states that little has been done to investigate the origin or early days of agriculture. "Darwin, de Candolle, Dr Pickering and the others had given the matter some attention but hardly treated it from an anthropological point of view." Few anthropological students had, he wrote, anything to do with agriculture. An appendix describes the Rothamsted experiment in continuous wheat growing without manure, something much more fully treated of in the Journal of the Royal Agricultural Society.

Prolific writers are the most difficult for a bibliographer to encompass if only because of the number of their works and the uncertainty of having not only discovered all of them, but also of having seen and appraised them. Such a man was John Scott, editor of the *Farming World* and before that Professor of Agriculture and Royal Economy at the Royal Agricultural College at Cirencester. This writer's first shot at the target of fame was very practical and indeed informative to the admiring reader. It was *Farm roads, fences and gates; a practical treatise on the roads, tramways and waterways of the farm; the principles of enclosures and the different kinds of fences, gates and stiles.* This was an 123 pp. 8vo, illustrated, published by Crosby Lockwood of London in 1883 and was one of Scott's Farm Engineering Textbooks, i.e. no. 3 of that series. Scott was pretty cynical about the condition of farm roads in general. "Good roads", he wrote,"are rarely met with on the farms . . ." Mostly there were no made roads. There are many roads over which there is much farm traffic at certain seasons of the year that seem to be cared for by nobody. The Vale of White Horse when I was a young cyclist had only gravel roads throughout the area (the same when I was a boy in the early 1900's!). Roads and fences do not contribute directly to the review of the farm and all very naturally to the expenses but they are works of indisputable utility. Scott supplies some quite astonishing costs e.g. for every acre of enclosed land in this country there is over £1.00 invested in fences. There are other cost analyses, but the book deals with all kinds of fences, roads, pavements and footpaths. Good roads, wrote Scott, are hardly met with on the farm. In an age of universal progress and advancement the roads of the farm have received no general attention. No. 6 of Scott's series was *Field Implements and Machines; a practical treatise on the varieties now in use with principles and details of construction, their points of excellence and management,* an illustrated 8vo of 181 pp. also issued by Crosby Lockwood in 1884. In that year another of Scott's *Farm Engineering Textbooks,* indeed the number one of the series, came out. It was *Drainage and Embarking: a practical treatise embodying the most recent experiences,* some 133 pp. 8vo with instructive illustrations. The year was full of Scott's industry. He added to his output *The soil of the farm*, no. 4 of *Handbooks of the Farm*, a series of which J. C. Morton was the general editor; it was an 134 pp. 8vo, a second edition of which was published by Bradbury Agnew of London in 1884 and a 5th undated issue later. In that year too Crosby

Lockwood published *Basic Barn Implements and Machines; a practical treatise on the application of power to the operation of agriculture,* No. 5 of Scott's series, an illustrated 8vo of 199 pp. Another was No. 4, *Farm Buildings, a practical treatise on the buildings necessary for various kinds of farms,* a 3rd edition of which came out so late as 1897, an illustrated 8vo of 167 pp. I am afraid I can do no more than supply details of these publications, whose general aim seems to have been to be successful, popular, more or less elementry productions. This will perhaps be enough to indicate their character and the public they aimed at.

A more technically informed and practically employed public was aimed at in two other books both about sheep farming. The first, *The practice of sheep farming,* 210 pp. with a frontispiece issued by T. C. Jack of Edinburgh. The second was a combined effort by John Scott and Charles Scott bearing the title *Black faced sheep; their history, distribution and improvement. With methods of management and treatment of their principal diseases,* an 8vo pamphlet of 307 pp. with a frontispiece, plates and other illustrations, also put out by T. C. Jack of Edinburgh and London in 1888. The book is dedicated to Charles Howatson, whose portrait forms the frontispiece. He was a famous breeder of these sheep whose origin is, he said, lost in obscurity. This is less important than that the breed has retained its purity from time immemorial, though he esteemed it much improved both in mutton and wool, which have been almost doubled within living memory — a suspicious statement not coincident with what Scott had already said. Scott supplies a list of breeders who had exhibited during the then past year, nearly five pages of names. Thereafter the book deals with every aspect of sheep breeding, care in lambing, sheep washing, dipping against disease, the value of different types of grazing, the improvement of hill pastures, sheep dogs, diseases, no phase of sheep breeding being omitted, a veritable encylopaedia of the subject.

Martin John Sutton was a seeds man who wrote a book. It was *Permanent and Temporary Pastures. With descriptions and coloured illustrations of leading natural grasses and clovers.* It was on, or better said an expansion of, an essay publication in the Journal of the Royal Agricultural Society, Vol. 22, pt. 2. It was a large 8vo of 158 pp. published by Hamilton Adams, London, 1886. This went into several editions. A second edition seems to have been put out in London, 1889, some 175 pp., illustrated with large coloured plates. I have not acquired a note of all the editions but a 5th of 204 pp. with

23 coloured plates was published in 1895, a remarkable popularity for such a recondite work. What is called a popular compilation of 144 pp., 8vo, was issued by Simpkin Marshall, London 1911. The structure of this book is rather what might be expected by the enthusiastic reader. It was a series of plates each of a single grass or clover annotated with a botanical description. It is reasonably complete but it would be difficult to say much more about this book comprehensive as it is. What I have said must inform the reader of what he might expect to learn from it. The subject was expanded by Sutton in *Botanical descriptions, analyses and illustrations of the principal grasses and clovers used in permanent pastures and in alternate husbandry. Reprinted from the popular edition of Permanent and Temporary Pastures,* with two sets of pagination, 129—175 pp., large 8vo, published by Hamilton Adams, London, 1888. The lengthy titles that were the fashion of the time give so clear an idea of the contents of the published books that little is left for the assiduous bibliographer to say — so I will refrain from further exegesis.

This is not of course true of all and every of the books of the period. One of these was written by one John Walker. It was *The Cow and the Calf in health and disease,* a small 8vo of 125 pp. issued by Thomas and Jack, 45 Ludgate Hill, London in 1886. There is an illustration of a cow opposite the preface which propounds that if civilised people were ever to lapse into the worship of animals, the cow would certainly be their chief Goddess. She is the mother of beef, source of butter, original cause of cheese, to say nothing of the shorthorn, hair combs and upper leather. A gentle amiable creation, ever yielding, who has no joy in her family affairs she does not share with man. We rob her of her children that we may rob her of her milk and only care for her when the robbing may be perpetuated in Household Works!

Walker was not precisely a modest man. In his preface he declared that having found no book devoted to the cow and calf "pure and simple", his small work was presented as being capable of being utilised by every cowkeeper. "The present counsel", he wrote, is given from actual experience. Some repetition may be observed in the manual but he believed each part might appear or was complete in itself. Walker was convinced that rearing and fattening of cattle was the job for the English farmer whose wheat yields could not compete with the reclaimed lands of the far away United States. "The pasture lands of England", he wrote, "when cared for will make the meat market". He then describes the contemporary state of various breeds, after which he went on to the Gentleman's cow, the Breeder's cow and so on, the milk sellers' cow, e.g. the Ayrshire. Besides the Ayrshire Walker recommends "The Cotters' Cow" which was a choice between three breeds, the Ayrshire, the Welsh and the Kerry. This author then proceeds in detail how to manage the animals in health and disease. It must have been a useful book in its currency in the 1880s.

I would not wish to engage in any sort of political controversy in these pages but I think I feel obliged to mention, if no more, Charles Brodbaugh's *Compulsory cultivation of land; what it means and why it ought to be enforced,* a 25 pp. 8vo put out by the Fruthought Publishing Co., London 1887. I say no more about it as I am not here engaged in a political polemic.

It is not often that a judge plunges into the thick of controversy about the polemics of the subject or its political implications. Judge Hastings Ingham was courageous enough to do this (see *The Country Gentleman's Magazine*, Dec. 1974). He wrote *Agriculture, its history, importance and prospects*, a paper bound 8vo of 112 pp. published by George Routledge and Sons of London, Glasgow and New York in 1887. "Unfortunately", he wrote, "even with all our boasted civilisation in England never has been the divinity far more worshipped than Ceres who instructed Triptolemus so of the king of Attica in everything pertaining to agriculture, and then gave him her chariot and commanded him to travel all over the world and to communicate his knowledge of agriculture to the rude people who had hitherto lived upon acorns and the routes of the earth". He thought it sad to have to say no victory of peace was ever so long delayed as the conquest of the soil. "Today I may perhaps comment that it is not only conquered but being destroyed for example the dust bowl in the United States". Ingham cites the Biblical story. Elisha was ploughing with 12 oxen (a team only seen years ago in the wildest highlands of Scotland) when Elisha passed by and cast his mantle upon him.

"Agriculture . . . farming was the first occupation of the first human being that trod this earth." Again "the history of agriculture is the history of civilisation and progress in the earliest times. In following the track of the ploughshare we shall have the satisfaction of tracing the rise of human industry from its very origin and of watching those early struggles with material Nature and those triumphs which skill and labour have accomplished". But quotation while appealing may become tedious. The book is most interesting and could be said to be well informed if not altogether scholarly. I cannot refrain from another quotation.

"I may here point out a remarkable difference rarely noticed (even today) though so palpably true which exists between man and the rest of the creation. The quadruped, the bird, the reptile, the insect and all the countless millions of creation have their food or means of subistence prepared for them. (Presumably by God!). To this law of providence man is the exception. He must labour in some way or other for his food. Not one acre of land in the universe spontaneously produces for man". But I would ask here — What of the food gatherers, some of the earliest men?

Of textbooks at any date of modern times there is always a plethora! One Prentice Manning produced a 32 pp. illustrated 8vo pamphlet, reprinted from the *Land Agents' Record* in 1887 but we could not include it here because we could not find a copy of this in the British Library. It is inserted here because it is listed at Southampton.

The appeal to authority is commonplace to writers of textbooks. How could it be otherwise? Samuel P. Preston was no exception to the justice of this simple observation! He wrote *Pasture Grasses and forage plants and their seeds, weeds and parasites,* an illustrated 8vo of 144 pp. published by T. C. Jack, London and Edinburgh in 1887. It seems to have been one of what was called The Cow and Calf series. In the preface to this book there is a great deal of discussion of the comparative merits of different grasses. "Others", he wrote, "advocate grasses etc., unknown to our agriculture". Preston had the best of intentions! He declared that he had written the work with the object of stating the essence of the British, Continental and American authorities. "Of original investigation into the subject of pasture and forage plants there has been a surprising dearth in the United Kingdom". Since Sinclair little or nothing has been done. There was one exception to this routine observation. I will add that of Sir J. B. Lawes at Rothamsted with which I can only suppose Preston was acquainted. He certainly should have been, though it does not immediately appear from this work. Preston gives a comprehensive list of authorities. "Grasses", he wrote, "are the true plebians of the vegetable kingdom", and he deals with them botanically and agriculturally in sections. In spite of my remark about his omission of the work at Rothamsted, I must assess this as a very able little book.

It was the fashion in Victorian times for a successful business or professional man to buy a country place, or estate, and to set up as a country gentleman. It is quite the same today. One Arthur Roland was a man of this genre! He was a common writer who wanted his children to live in the country, so he took up part-time farming. He set up with 40 acres of grass and plantations, 10 acres of arable and 5 acres of hops. He declares that his business friends persuaded him to become an author, but I can only suppose that he was infected with the virus which assails all individuals who would be writers. In the result he put out no less than 8 vols. of instructive works all of which were published by Chapman and Hall and came out between 1879 and 1881. I think little more need be said about them than the details of their titles and so on because they are self-explanatory and the books do not add to the then sum total of knowledge for farmers, or possibly better said textbooks that might tell them something about the job or jobs they had known since childhood!

In the British Library these books are catalogued as Arthur Roland, *Farming for pleasure and profit,* 8 vol. The first of the series (by date) is *Poultry Keeping,* a small 8vo of 162 pp. It opens with an account of ancient domesticated poultry, deals with egg imports and with poultry keeping in towns and suburbs, indeed every aspect of the subject from the egg to the plucking of mature birds for the table.

The next volume to appear was *Dairy farming management of cows* etc., another small 8vo of some 250 pp. It was like the others in the series edited by William H. Ablett. The present work was suggested by inexperienced friends, or so said the author, who makes some rather interesting references to Jethro Tull, Arthur Young and others of his predecessors. The various breeds of cows are described, and so is the dairy and its untensils, calving, the rearing of calves and so on.

The third section was a small 8vo on *Tree planting* of some 157 pp., which need not, I think, detain us, as the subject while analogous to farming is not precisely farming.

The fourth section, which was put out in 1880, was another small 8vo of 194 pp. with the title *Stock keeping and cattle rearing.* Roland describes the various breeds, the points of good animals and their selection, oxen, sheep and pigs. This little book also deals with the diseases of pigs, oxen and sheep.

The fifth part of this series which was published in 1880 was *The drainage of land, irrigation and manures*, another small 8vo of 192 pp. The details include a plan for laying land down to grass, as well as the best method of the preservation of manure (by manure is meant the animal faeces) dropped in stall or stable.

The sixth section allied two crops which seem to me to be quite different in physical appearance and indeed methods of production, but one cannot today reason with an author long since dead. It was *Root growing and the cultivation of hops*, an 8vo of 184 pp. This volume opens with a chapter on the agricultural labourer and suggests that he should keep pigs and cows, the former perhaps a possibility, the latter an unlikely enterprise for that class of worker. It also discussed the *Report on the employment of women and children in agriculture*. The skill of the workers is highly esteemed but the habit of giving the men beer is condemned. Roland then proceeds to the proper subject of his title, the cultivation of turnips and other roots including potatoes. The potato disease, the handicap of that time, is handled and the use of various manures, with various theories of the diseases. This small volume is completed with instructions for hop growing.

The 8th section of this series was *Market garden husbandry for farmers and general cultivations*, another small 8vo of 187 pp. dated 1881. It deals with the cultivation of various plants too numerous to enumerate here, including the always popular and possibly ever present potato.

This series was completed by another small 8vo dated 1881 on *The management of grassland, laying down to grass, the artificial grasses etc.* It was provided with illustrations of the various grasses, and the botany of grasses. The conversion of arable to pasture, a frequent operation of that day, is detailed and there is some description of experiments made with manures.

Frank Humphrey Storer was a professor of Agricultural Chemistry at Harvard University and wrote a two volume 8vo of 529 and 507 pp. respectively. It was a large and exhaustive work and was published in London by Sampson Low and Co. For such an exhaustive treatise it had a remarkable life. It went into no less then seven editions. This author must have been at least adequately rewarded. The preface made the pretence that "This book has been written in the interest of persons fond of rural affairs", which seems to disclaim any intention of instructing the practising farmer, who would in the majority have little time to spend in making such a massive work but this condition is confronted with the author's protests that the book consisted of lectures much amended first given or possibly intended for young farmers and the sons of farmers familiar with the manual practice of agricultural operations, and also to men who intended to establish themselves on farms or occupy country seats or become landscape gardeners. The book is very scientific and could not be discussed in detail here, not only because of its size but because it is largely American, though agricultural chemistry has a great deal of common application in the two countries. It covers the whole range of possible manures, animal, man and vegetable.

Chemistry is a vast subject and is the care of the diligent farmer, but a two volume work of nearly a thousand pages is rather more, or so I should persuade myself, than the most literate of Victorian farmers would care to peruse. It must have been restricted to such shelves as the unreal books with which the farmhouse may have been provided. It was, as already said, dealt with *in extenso*. Frank Humphreys Storer's book was issued with the title *Agriculture in some of its relations with chemistry*. Vol. I, pp. 529; Vol. II, pp. 495. The book was published by Sampson Low, an 8vo.

In the preface the author protested that his book made no special appeal to chemists or students of chemistry. It was based on lectures given during sixteen years at the Bussey Institute (what and where was it?) and many times altered and revised. The two volumes are carefully detailed with necessary tables. Storer acknowledged the help of his teachers and the sources which he had used.

In his first page Storer asks a question which has not yet been fully answered, though it has in some part. He wrote, considering the relations in which plants stand to the air and soil questions naturally arise: "What are their food sources?" and he asked the question "How do plants take in their food?" Leaves take in carbonaceous matter formed by carbonic acid. From the air too comes the main part of oxygen, the predominant constituent of the dry matter in plants. A fresh or living plant, he wrote, largely consists of water. He then describes water culture in which plants grow tolerably well. Bonnet likewise studied it long ago. As early as 1758, Duhamel, a French chemist and botanist, grew beans, chestnuts, oak and almond trees.

Storer seems to have made some sort of experiments. He discussed "Sand culture", and had apparently made some experiments growing plants in jars filled with pulverised quartz plus water. He said that "In Holland they use the method for bulbs on the sand dunes", but his remarks on this subject are far from clear. Indeed these two volumes are so detailed that it is probable that any farmer who owned them would just study that part which applied to his farm, for it would be a large task to digest the whole.

That times are bad is no new complaint: it is as a matter of fact continuous and dully repetitive. One Col. Tabor joined the ranks of this class of grumbler with a pamphlet of 12 pp. 8vo published in or near about 1887. Its title was *Small farms* . . . His moan is quite usual. We are not as we were. Order has given place to disorder, religion to unbelief, virtue to vice and respect for the persons and property of others to an unscrupulous socialism. The middle, the backbone of England is fast disappearing. This little pamphlet is really a plea for small farms. "In old times" wrote Tabor, "men were not accustomed to speak of land by the acre but by the plough". No more need be said of this minor effort.

John Walker, of the Hill near Rugby, was a quite different type. I can only suppose he was a successful farmer. He was certainly a voluminous writer and work from his hand was published after 1900, the date which marks my limit, but does not necessarily prescribe a limit to a writer's life or his activities (which is of course self evident!) The earliest of his productions appears to have been (I am a little uncertain about this) *The Cow and the Calf: a practical manual. With a description of the various breeds of beasts, their milking capabilities, dairy work and an article on the baneful ergot parasite in grasses*, some 128 pp. provided with a frontispiece and other illustrations, only the second edition of which I have seen. It was published by T. C. Jack of London and Edinburgh in (subject to a query!) 1888. The same subject was also treated of by Walker in *Cows and the dairy*, 146 pp. 8vo edited by T. W. Sanders and published as No 8 of a series, *Profitable Farming Handbooks*. Both these books are a compendium of the treated subject, which it would be tedious to expand exhaustively here. The same could be said about *The Sheep and lamb: a practical manual on the sheep and lamb in health and disease*, an 8vo of 135 pp. issued by Thomas C. Jack, London and Edinburgh. The preface opens with some enthusiasm for the subject. "In resuming the Cow and Calf series" the sheep and lamb take the next place, its preface being enthusiastic about the subject: "both in ancient and modern times the farmer recognised in his sheep flock his greatest source of profit . . ." (but) there was still room for improvement (there is little in human affairs of which this is not true!) and he becomes rather trite when he remarks that an ounce of practice is worth a pound of theory. The diseases of sheep have been neglected, he declares, and proceeds to inscribe a large section on the treatment of such ills. The book itself opens with the history of the sheep. The Abel of the Bible was a keeper of sheep. Ewes, he wrote, love their offspring. He describes the various breeds. Illustrations of ewes expand the book and there is, rather curiously, I think, one of ergot in grasses.

Another production of this writer was *Farming to profit in modern times, arable and pasture lands including Preparation of land, sowing and harvesting, breeding and rearing of stock, dairy etc with a Calendar of Work laid out for each month*, an illustrated 8vo of 136 pp. put out by Thomas C. Jack of London and Edinburgh, undated but possibly 1888. The title is so expansive that little need be said of its contents, which were strictly orthodox.

This author went on writing into the 20th century, but his later books need no more comment here, i.e. they are beyond my self appointed time limit, except perhaps a mention of two more titles, the first, *How to farm with profit arable and pasture land. A practical manual on modern agriculture*, and *Pigs for profit*, which was edited by T.W. Sanders in 1910, a 110 pp. 8vo.

One Thomas Walley was, it seems, more of a veterinarian than an agricultural writer. An Edinburgh 4to was entitled *The four bovine Scourges*; dated 1879, it is obviously, or so it seems to me, a veterinary book. *A practical guide to meat inspection* is not likely to be of any agricultural consquence, although it was rewritten and enlarged by Stewart Stockman in 1909 (?). His pamphlet bearing the title *Hints on the breeding and rearing of farm animals*, a 68 pp. 8vo put out by Turnbull and Spears of Edinburgh is not to be found in the British Library, at least by me.

It must always be gratifying to an expert to be asked to write about his job. Such a one or so he declared was Robert William Ashburner of the Manor House, Moreton, Warwickshire and appropriately enough for a resident in that county his subject was the Shorthorn. His first gage with which to challenge the house of fame was *Shorthorn experiences written at the request of a few breeders*. It was a 207 pp. 8vo provided with a frontispiece and other illustrations published by H. T. Cooke of Warwick and Simpkin Marshall of London in 1888. He said himself "The author of this small book has frequently been requested by several friends and shorthorn breeders to write a short account of his proceedings," a very ordinary protestation by a rather moralising man. On p. 6 he said it is just over 100 years since Collings brought his first Duchess cow and Hubback and followed this remark by some herdbook-like information which I need not perhaps expand here but it will be well to say that some of his animals were described from the breed of Bates of Kirklevington. Part II of the book is the author's life and experiences. In North Lancs where he was born "The inhabitants then knew nothing of the bustle and hurry of the present day" (how frequently has this been subscribed to at all times?). It was a purely agricultural district, no travelling by rail, nor were the postman's knocks to be heard at the door, but by the time he was writing it had become a large and prosperous town.

This book is of course largely a discussion of breeding practices followed by him, coupled with his arguments in support of his own methods and some odd anecdotes and moralising.

Ashburner also produced an 8vo of 460 pp. with the title *The Shorthorn Herds of England 1885-6-7*, which was published by Cooke of Warwick and Simpkin Marshall in the same year. This book is really a description of all the Shorthorn herds then in England, a very detailed directory supposed to supplant Dixon's *Saddle and Sirloin* of some twenty years earlier. It does not lend itself to discussion but is a useful and detailed piece of historical reference.

One William Day was a writer about horses, the turf and kindred subjects. One largish production was his *The horse how to breed and rear him*, a large 8vo of some 453 pp. published by Richard Bentley and Son, London. Chapter XXI has the caption "The half bred on the farm", but Day thought the farmers ought to use mares. They could, he said, work until a little before dropping a calf. Thoroughbred stallions were very good, perhaps a self evident observation. A stud of 24 mares was the optimium size in his opinion. He could not set out the working hours of horses because practice varied widely in different counties as did the men's wages. Carters often worked thirteen hours a day. A visit should be made to the stable between 7 and 8 p.m. to see that all was well. Ploughing was in his opinion generally too slow. He thought four ploughs could well be used on a 3 acre field i.e. 3/4 acre each. He makes various exordiums on management and work and expands the subject in the last several chapters. This must have been a useful book for any farmer who read it.

Henry Kains Jackson edited *The practical guide for making ensilage in stocks and silos*, which was issued by the Ensilage Society (of which I know no more than the title) a 50 pp. illustrated 8vo published by Eyre and Spottiswood, London, in 1888. Before this he had written *The fields of Great Britain,* an 8vo of 1881 which is all I know about it. In 1885 his *Experiments in making ensilage during the wet season of that year,* a 60 pp. 8vo was edited by E. C. i.e. Ernest Clarke. This is only a list of his contributions, but their titles are, in my opinion, sufficient here where it is not intended to write a history of farming.

I must be mistaken, or so I think, because I could find no entry in the British Library Catalogue setting out the work of Eleanor Anne Omerod but the Royal Agricultural Society has (or had) an apparently complete set of her works; a note of the titles and other detail is appended.

The earliest of these useful productions is (or so I think) *A manual of injurious insects with methods of prevention and remedy for their attacks on food crops, forest trees and fruits,* a 410 pp. 8vo supplied with a frontispiece and other illustrations, issued by Simpkin Marshall in 1881, a second edition of which came out in 1890. In 1888 a 4 p. illustrated leaflet *Notes on the warble fly or bot fly* was printed by West Newman, London. Next came *Guide to methods of insect life and prevention and remedy of insect ravage,* a London 8vo of 1884. This was followed by *Notes and descriptions of a few injurious farm and fruit insects of South Africa with descriptions and identifications of the insects by O. E. Janson,* a London 8vo of 1889. Next came *A textbook of agricultural entomology being to the methods of insect life and means of prevention of insect ravage,* the second illustrated edition of which was issued by Simpkin Marshall, some 258 pp. 8vo in 1892. The next was a *Handbook of insects injurious to orchard and bush fruits with means of prevention and remedy,* a London 8vo of 1898. Finally for my present purpose was *Flies injurious to stock. Being life histories and means of prevention of a few kinds commonly injurious with special observations on the warble or bot fly,* an illustrated 8vo issued by Simpkin Marshall in 1901.

The Royal Agricultural Society Library Catalogue lists a work by H. M. Upton, *Profitable Dairy Farming,* a London 8vo of 1888 but we could not find this in the British Library Catalogue. Alas!

In 1888 one John Walker wrote a small 8vo of 129 pp. which was pubished by Thomas E. Jack, London. Its title was *Farming to profit in modern times, arable and pasture lands.* The writer was already a recognised authority, for he modestly states in his preface that this book the author was encouraged to write as the public approved a

"Cow and Calf" and "Sheep and Lamb", his previous works. The notes herein contained, he writes, must not be needlessly followed in so variable a climate. With fair allowance for seasons and careful calculations they should be found useful. No man should be utterly the slave of custom or content with the beaten path.

In this book crops are discussed in detail as well as meadow land and its treatment, animals and their needs and the capital required. The use and care of machinery is laid out. Walker was acquainted with the literature and was also an enthusiast for the growth of sugar beet, today so important a crop.

In the same year W. H. Wheeler produced a 168 pp. illustrated 8vo with the title *Drainage of the Fens and Lowlands by gravitation and steam power,* issued by E. and F. N. Spon of London and New York. In his introduction Wheeler laid out his idea in producing the book. "Primarily this book," he wrote, "is a general description of works and machines plus practical limits and information that may be of service to those superintending drainage districts or designing new works. A civil engineer has not always necessary experience needed to design details of the machinery required but able to judge the type best suited for the work required and draw up specifications and tenders to others. Insufficient knowledge of the special subject has led to unsuitable machinery and waste of money without fulfilling the object required."

This is certainy detailed in every way, inclination of surface water, slope of drains, area of land to be occupied by drains and means such as windmills in Holland. Cleaning and removal of weeds is carefully described with different methods used in different areas. This is a book of great value to the folk involved at that time with a wealth of details and illustrated plates. It must also be of some value to present day students interested in agricultural history.

The Royal Agricultural Society has in addition *A history of the Fens of South Lincolnshire,* 2nd ed. greatly enlarged, a Boston 8vo of 1894 (?). This is not, however, a farming book.

As much or as little could be said of Theodore Wood who seems to have been a naturalist. He wrote an illustrated 8vo of 396 pp. over the title *The farmers' friends and foes,* published by Swan Sonnenschein in 1888. It is economical with the pictures which are small plates in the text, not separately as was sometimes or often usual. Wood modestly said in his preface that he had endeavoured not merely to give an account of the various animals which influence British agriculture, but to show the intimate connection between the injurious class and the beneficial, and the fallacy depending upon our own efforts rather than upon those of Nature in the unceasing struggle with our natural enemies. His thesis is really that Nature controls Nature, although he does not say so.

He does say that in entering upon the great and all important task of Agriculture without which mans' existence in the world could be little better than a pure and simple savage (are they pure and simple?) we are confronted with the hostility of Nature whose provisions we are absoluty compelled to disobey.

The book contains a lot of Victorian philosophy about the relation between animals and man. The book is in two parts: I, foes (of man), rats and mice, birds and insects. The friends are the beneficial birds and insects. It is completed with a brief but caustic summary.

Any author who would preface his writing with the statement that a want has long been felt — by what of course he does not say — could doubtless specify this to his wanting public. Such was the pretension of Archibald N. Mc Alpine, BSc London, who wrote an informative, or perhaps instructive would be a better adjective, book with the title *How to know grasses by the leaves*, which was distinguished by an introduction written by Robert Wallace. It was a sufficiently illustrated and descriptive work of pp. 3-92, plates, 8vo, issued by D. Douglas, Edinburgh. The preface modestly remarks that "a want has long been felt of some simple and attractive guide to the identification of the common pasture grasses by the leaves". This vacant line of learning he proposed to fill up. "The inferior grasses were" he wrote "neglected by stock and so flourish and are allowed to mature their seeds." The contents are complete in a small space and McAlpine was wise enough to advise that grasses must be studied throughout a whole season. Wallace in his introduction praises the work. The illustrations are detailed and informative, the grasses classed by colour and appearance during growth.

Besides this original work McAlpine also translated *The best forage plants* from the German of Friedrich Gottlieb and Carl Schvoter, dated 1889. The plants are fully described and figured with a couple to account of their cultivation, economic value, inpurities and adulterants and so on. The translation made by McAlpine was a large 4to, 171 pp., illustrated by plates. No more than this need perhaps be said here about this translation.

An author named William Day wrote a book of which it seems to that the title is a sufficient indication of its contents. This was *The horse: how to breed and rear him — race horse, hunter, hack, troop horse, draught horse, pony*, an 8vo of 453 pp., the second edition of which was put out by R. Bentley of London in 1890. A description of the contents could be expanded here, but is rather to much for the immediate purpose.

An American author who made a name of himself and whose theories were largely expanded in this country has a rather prescriptive right to be included in this ambitious and hopefully complete work — of its sort! Wilkes P. Hazard was a citizen of the United States but he must have had a public in Great Britain as well because his work was published here by Trubrier and Co., an 8vo of 144 pp. with a portrait as frontispiece and nearly 100 illustrations. The title was *How to select cows on the Guerion system simplified and explained and practically applied.* The illustrations are very explicit and show up the characters very well. The system is however very complicated and I cannot assume that it was widely followed in this country where it was, I believe, very largely theory and very little practice.

Just the opposite may be said of Henry F. Moore's essay on *Royal Agriculture at royal Windsor. Description of the royal farms at Windsor; history of the Royal Agricultural Society; and guide to the Windsor Show*, an 8vo of 134 pp. published by Bell's Weekly Messenger Office in 1889.

As might be expected, the author offers his humble thanks to her Majesty for her gracious permission to view the farm and for permission to dedicate this work to her Majesty. Chalmers Morton wrote, there is no department of British industry which has made a steadier progress and success than our Agriculture, which has achieved more obvious improvement during the happy reign of Queen Victoria. What was true in 1860 is doubly true now, with the evidence of the Royal Agricultural Society of England during her reign. The model has been uniformly set by the great landowners in no small measure due to the example set them by England's Royal Family. As Horace put it, the possession of land has always been among the highest ambitions of men. It was remarked by Horace:

This used to be my wish,
A house and garden with a spring at hand
And just a little wood

Kings, wrote Moore, have played a great part in improving their lands. Irrigation by these folk caused inundation and irrigation affecting Assyria, Babylonia, Chaldea, Egypt and Palestine. The numerous canals from the Tigris and Euphrates were called for and so the different forms of agriculture were fostered.

History tells us that in ancient times agriculture was beloved by royalty.

Enough, Moore went on, has been said of the great practical interest shown by rulers. Hence the calling may be regarded as an honourable one and the influence of the Royal Family is shown in this book. This author acknowledges that the facts given in this article are from John Chalmers Morton's *Memoir* of 1863.

The Show Farm stands first with a magnificent model of buildings. It is also the Home Farm of the historic home of the sovereigns and includes the magnificent Royal Dairy which had recently been remodelled and modernised, as the term was then understood, the changes being specified and carefully described. There was also the then famous Flemish Farm, comprising both arable and pasture. Details of the machinery installed and the cattle kept there follow. Good crops follow good management.

This book brings home to the reader the solid foundations of our agriculture. The Royal Patronage enjoyed by our countryside gave it the necessary impetus which has been carried on ever since in all its branches.

Lecturers are apt to write books. Charles R. Valentine of the Devon Cow Dairy Farm, Ludlow, was one who did this as well as keeping a dairy farm of his own. His job was lecturer and practical demonstrator on dairying at the University College of North Wales. His book, *Buttermaking*, a 76 pp. 8vo, was provided with a variety of illustrations. The book, which is dedicated to the Earl and Countess of Aberdeen, is based on a lecture given at the Agricultural Dairy Conference and Show at Ludlow in 1888, which Valentine had been pressed to publish as dairying was then of growing importanace. The introduction contains tables of butter imports (p. 10) and margarine imports (p. 11). The imports gained a market because they were of uniform colour and quality but should not be encouraged "where we can make our own", a subject discussed at length. The chapter on feeding mentions ensilage (then a burning topic!) which Valentine said was good for dairy cows. Other feeds he mentions are maize, malt dust, wheat bran, and brewers' grains. The illustrations explain the methods discussed. The Laval and the Danish separators are illustrated and described. This seems to be, or possibly I should say is, a book that expounds details of the whole of the dairy industry from cow keeping to milk separating and butter making and so on. The illustrations are most informative. A curiosity of the book is that it contains advertisements in Welsh.

William Wilson Jr. was by no means so ambitious, although he was a Scot from Aberdeen. In 1889 he published *Practical Observations on agricultural grasses,* an 8vo of 117 pp., put out by Simpkin Marshall. He more or less apologised for this effort by saying in his preface that in the eyes of agriculturists it is held indispensable that any investigator should posses a practical knowledge of agriculture and the habits of agricultural animals as well as a scientific knowledge of the plants most likely to be useful. The success of the work has surpassed his expectations. There is a large frontispiece with pictures of various grasses. In the text Wilson states his idea of the then present position of agriculturists and makes a good deal of criticism of farmers who had not, he said, a thorough knowledge of the various grasses, as well as criticism of the poor quality of seeds supplied to farmers. He gives details of the varieties of rye grass and discuss grasses in general, giving details of them all including Lucern and Sainfoin. He instructs his readers in how to use the grasses, giving seeding rates and the suitable soils for each and ends his book with a second plate showing the grasses.

Wilson also wrote *Investigations into applied nature,* a London 8vo of 143 pp., dated 1896, a second edition of which came out in that same year. It is perhaps too recondite to include in detail here, a mention of it being, at least in my opinion, sufficient.

Far be it from me to exaggerate the difficulties of doing this work for it is my own choice to do it with always the encouragement and most capable assistance of my wife, but it is not without its disappointments. One of these is the apparent disappearance of a book by one F. J. Lloyd, entitled *The Science of Agriculture,* published in 1854, which cannot at this present moment (October 1985) be found upon the shelves of the British Library, so for the present purpose it must be written off as a loss.

William Fream was a voluminous writer with a large output between 1888 and the end of the nineteenth century and even later so I shall not pretend that the following list and conmments is absoluty exhaustive.

The earliest venture into print made by this prolific writer was published by H. Cox, The Field, London in 1888. It was *The Rothamsted Experiments on the growth of wheat, barley and the usual herbage of grassland*, an 8vo of 235 pp. The title is sufficiently descriptive, and it would be a work of supererogation to expand upon its contents in this place. The next work of this industrious author to come off the press was *Soils and their properties*, one of Bell's Agricultural Series, an illustrated 8vo of some 176 pp. dated 1890. It is unusually exhaustive for a work published in a series of this kind and is almost frighteningly modern in its contents — than which perhaps no more need be said.

Fream's next work was *Elements of agriculture*, a textbook prepared under the authority of the Royal Agricultural Society whose catalogue lists no less than ten editions, the last of which was a small 8vo, dated 1918, rather outside the scope of the present study but necessarily mentioned. The preface to the first edition modestly stated that the book was an elementary work on agriculture adapted for use in rural and other schools and classes. A set committee of the Royal Agricultural Society gave this job to Fream and a variety of pundits looked though the proposed work. Various editions of this work are filed in the Society's Library (or were when I spent some time working there a good may years ago).

The book was later translated into German by Dr W. Graf Gocotz-Wristerg and published as an 8vo at Berlin in 1893.

The Highland and Agricultural Society has also *The Complete Grazier and Farmers' and Cattle Breeder's Assistant*, rewritten and revised by William Fream and William E. Bear, an illustrated work in four volumes, published in London, no date being given. This I have not seen.

Arthur Bower Griffiths was a contributor to Bell's Agricultural Series with *The diseases of crops and their remedies: a handbook of economic biology for farmers and students*, an illustrated 8vo of 174 pp. put out in 1890. In the previous year he had contributed a volume to the same series *Manures and their uses, a handbook for farmers and students*, an illustrated 8vo of 159 pp. In 1889 too Griffiths wrote *A treatise on manures or the philosophy of manuring* one of a series known as the Specialists' Series, an illustrated 8vo of 399 pp.

In his preface to the last book Griffiths said that the main object of the volume was to detail in a concrete form the life spell of the principal insects and vegetable foes of the farm and give an account of the means of destroying them or of preventing them from attacking. The matter is very appropriately arranged under four heading:- (a) Diseases of leguminous crops (b) . . . of root crops (c) gramineous crops (d) miscellaneous.

The majority accept the tenets of their day. Griffiths was no exception to this rather arbitrary rule. He accepted what was then the accepted canon about antiquity. His profound reflections on the subject are that "The production of food as other articles for the use of mankind and domestic animals is one of the oldest occupations of mankind. Like the science of chemistry, medicine and leading generally soil conditions originated in the East." Cuttings representing figs and the fig tree are found on the Egyptian pyramid at Gizah. "It was very ancient, probably 1,500 years B.C., while the date of its erection was about 34 centuries ago. To build this necessitated a certain standard of civilisation and so an established form of agricultural existence", but he need not here be followed throughout his historical notes, which run from Fitzherbert to his own day.

This rather lengthy book might be described as a miniature encyclopaedia of agriculture and as such may perhaps be left here. Incidentally this author's works ran into the twentieth cecntury and may perhaps be left here.

Towards the end of the 19th century several publishers put out what might be called textbooks or books of instruction for both the students of the South Kensington and provincial institutions that were set up to instruct either young aspirants to the farming profession or established farmers who wished to know of both the practice and the theory of their job. Amongst such publications must be remembered The Shilling Warne's Useful Books. For this series one George Armitage, MRCVS, wrote at least three books in 1890. *The horse, its varieties and management in health and disease*, an 8vo of 259 pp. with a frontispiece showing The riding party.

Armitage was at the very least an honest writer. His book was, he said, condensed from Stonehenge's *British Rural Sports, The Farmers' Calendar,* the *Modern Householder* and other works. The book is quite comprehensive, a *multem in parvo*. It opens with the early history of the horse but he admits that this early history is wrapped in obscurity and fable, and little or nothing is really known of it except that we have reason to believe that he first came from Asia, like man and according to the Mosaic account all other animals now existing, and that he was used in Egypt more than 1600 years before Christ. But with the history of the horse, he protests, he shall not encumber this book. I surmise he did not know it.

The habits of the horse, he wrote, in all countries and of all varieties are pretty much alike. Wherever he is at large he is bold but wary. There were wild horses then in the interior of Asia and in South America. But both the horses of the Tartars and those of the Plate are descended from domesticated animals and could scarcely be called wild. California horses which have still more recently been bred in a wild state from Spanish horses are quite as those described by F. B. Head Kt. Bt. in *The horse and its rider*, 2nd ed., a London 8vo of 1861 — strong and durable.

The then present varieties of the horse were the Arabian Barb, an African horse, Dongola horse, another African variety etc., etc. in expansive mood. The farm horses of his day, or so he said, were of three main breeds, the Black Dray, Clydesdale, and the Suffolk Punch; the Cleveland was mostly a carriage horse.

He proceeds to stables and management, feeding, breeding and diseases, accidents such as sprains and so on.

Armitage also wrote for this series *Cattle and their management in health and disease* and *The sheep, Its varieties and management in health and disease*, another illustrated work. Armitage was an indefatigable writer and extremely (in my opinion) productive but I do not propose to write any more about him in this place.

A Frenchman's book was translated by William Crookes. Its title was *The perplexed farmer. How is he to meet alien competition.* It was a translation of three lectures given to the Belgian Royal Central Society of Agriculture, Brussels, an 8vo of some 208 pp., put out by Longmans Green of London in 1891. The purely technical part was said by Crookes to be an admirably clear and luminous summary of the teachings of M. Ville's larger work. Artificial manures (p. 90) contains much additional practical evidence on the successful working of the author's system of artificial manuring. Ville writes on political measures, but much of what he said was totally inapplicable to English conditions. The second lecture gives an analysis of vegetables in their organic and mineral elements.

The second lecture is on the Use of Manures and contains various tables and diagrams. Of course the theory adumbrated is that of the then current day.

One A. W. Cruikshank produced a small 8vo of 29 pp. with illustrations published in Edinburgh in 1891 entitled *The dairy and butter making; a handbook for beginners*, which I have not seen but which could hardly have been exhaustive written in the compass of 29 pp. I can say no more of it.

Thomas Edward Kebbel MA, Barrister of Law, wrote an interesting solid study over the title, *The old and the new English Country Life: the country clergy, the country gentleman; the farmers; the peasantry; the eighteenth century*, an 8vo of 234 pp. issued by William Blackwood of Edinburgh and London in 1891.

In a rather apologetic preface Kebbel wrote "Some parts of this volume are only the reproduction in other words, of thoughts, opinions and descriptions first printed many years ago while others have been written with more direct reference to passing wants and the changed conditions of country society which I have lived to witness" (what aged man or woman has not?). He protests his acquaintance with the people who are his subjects, their habits characters and ideas . . . Were any one to write a story of English rural life entitled Sixty Years Since he would have to describe a state of manners almost as unfamiliar to the present generation as the manners drawn by Fielding and Richardson, George Eliot's earlier novels, Adam Bede, and Silas Marner. Scenes of clerical life have to some extent done this. But they belong to a still earlier period, the last ten years of the 18th century and the first ten years of the 19th, when not a ripple yet moved over the surface of rural life to tell of a coming change of weather (how wrong he was!). Villages clung to the old order of things.

Change was however then apparent in the character of the peasantry as well as the classes above them. These sketches are of a cast of characters more loquacious than those of Overbury, and possibly more profound. The last chapter dicusses the 18th century — the great men of that time — a period of tranquillity in the countryside and the continuation of old customs, a statement which is certainly not true.

Kebbel also wrote *The agricultural labourer, a short summary of his position*, 8vo, pp. 271, put out in 1870 by W. H. Allen and Co. Several later editions followed, the whole being founded upon the official *Report on the employment of women and children in agriculture* and his own thoughts. A second edition appeared 17 years later, in the introduction to which Kebbel said "Life however in the end of the 19th century moves slowly in an English village".

He refers to the Education Act of 1870, the Agricultural Union, etc. He saw the cure of the labourers' plight in a permanent and substantial rise in wages, but this work is about the labourer and not about farming. It was well received and went into several editions, the 4th of which came out in 1907, which is as much as need be said here. Kebbel was the author of several other books but these were not about farming or even about the labourer.

H. J. Mirehouse, according to the Perkins catalogue, wrote a 20 pp. pamphlet with the title *Sugar beet growing in England as a profitable crop* issued by the Times and Mirror Office Bristol in 1891 but we could not find this in the British Library Catalogue and so have not seen it.

Another doubtful work only to be found mentioned in the Southampton catalogue is said to be by William T. Lawrence, bearing the title *Agricultural elementary course. Adapted to the syllabus of the South Kensington Science Dept.*, an illustrated 8vo of 184 pp. published by W. and R. Chambers of London and Edinburgh in 1891.

A most prolific writer of the 1890s and later was Walter James Malden. Indeed I have decided to omit some of his 20th century writings. He was a contributor to several of the agricultural series published by various firms, one being Bell's Agricultural Series. To this he contributed *Tillage*, an illustrated 8vo of 156 pp. in 1891. To another series *Farm Field and Fireside Series* Malden added *Rational Pig Keeping to ensure profit*, No 1, a pp. 132 illustrated 8vo undated but c. 1893. In that year too W. A. May of the *Mark Lane Express* issued Malden's *The potato in field and garden*, a 217 pp. illustrated 8vo, the contents of which were highly instructive.

In 1896 Malden added to his quota of instuction in the *Agricultural Series* issued by Kegan Paul a work bearing the title *Farm Buildings and economical agricultural appliances*, an illustrated 8vo of 192 pp., the title of which is a precise indication of its contents, chapters V and VI being of special interest since they deal with machinery and give a view of the state of farm mechanics then.

His next contribution to knowledge was to produce another of Kegan Paul's *Agricultural Series* i.e. *The conversion of arable land to pasture*, which indicated that he was very much on the ball, for this subject was of burning importance at that moment (and indeed for some years thereafter). It was an 8vo of 190 pp. and I think its title is a sufficient indication of its contents to any reader who may be studying the period.

Malden also produced what seems to be a rather important work *Sheep raising and shepherding; a handbook of sheep farming*, some 160 pp. 8vo, provided with plates and other illustrations. In his preface to this work Malden exalts the sheep, which he said was aptly termed the farmer's anchor, which in spite of foreign competition it still remains and is likely to do so, the key to successful farming over the greater part of the country. Trustworthy books describing management of Scottish sheep are plentiful; books on English sheep management are few. Therefore this book is given up to English methods. Every breed and every district require different methods of treatment. This book deals with them broadly, while entering into the details of management by the best authorities as fully as possible.

In the introduction was outlined the skill needed by a sheep farmer. When this is ensured sheep farming is generally one of the most profitable sections of the farm. In case of frost the sheep farmer must look well ahead to crops for feed. This book is full of wise saws and modern instances, and could be largely quoted, but that would be invidious if Malden is to be compared with other writers on the subject. He made the rather obvious observation that the choice of ram and ewes was important and so was the cutting of the ewes.

The book gives a modern reader a clear idea of the ramifications of sheep farming. The farmer was a midwife and to a certain extent a vet; as well as a cultivator of land and a judge of animals for wool or for meat plus the correct way of feeding. These bygone farmers were most certainly dedicated men. The illustrations to this book are instructive and emphasise the different appearances of the main breeds. The few diagrams are plain and helpful.

Charles Morton Aikman's first aim at the "bubble reputation" was made in 1881 when Blackwoods published his *Elements of Agricultural Chemistry*, one of the Farmers' Club Library, but it was not until a decade later that he really got into his stride, beginning with a 65 pp. 8vo entitled *Farmyard Manure; its nature, composition and treatment*, published by Blackwood in 1892. This was in substance a chapter from a larger work on *Soils and Manures* on which he said he was then presently engaged. In this he proclaimed that farmyard manure was originally formed from vegetable substances and therefore contained all the elements present in the plant itself but what was absorbed in the animal body. He then proceeds to estimate the manurial value of the defecations of the various animals and goes on to a chapter on the action of farmyard manure on soils, an estimate of composition and effect.

In the same year an edition of James W. Johnston's *Catechism of Agricultural Chemistry*, revised by C. W. Aikman, an 8vo of 96 pp., was issued by Blackwood of Edinburgh. I assume that this needs no evaluation of its contents.

A much larger work, *Manures and the principles of manuring*, an 8vo of 592 pp., was issued by Blackwood in 1894. This book was dedicated to Lawes and Gilbert. It opens with an historial introduction, although "Agricultural Chemistry may be said to be entirely of modern growth," a catalogue of his predecessors in the field is supplied in rather overwhelming detail. This is an exhaustive treatise and sets out the then present state of knowledge and its uses. The author was then Professor of Chemistry at Glasgow Veterinary College.

An elementary pamphlet of 84 pp. 8vo was published by Vinton in the following year. It's title was *The food of crops and how to apply it,* described as an elementary handbook on the science and practice of manuring, which is perhaps all that need be said of it here, the subject matter being more or less obvious.

In the same year Aikman put out an illustrated 8vo under the title *Milk, its nature and composition. A handbook of the chemistry and bacteriology of milk, butter and cheese.* This was published by A. and C. Black of London. *Mea culpa*: I have no notes of the contents of this book.

Another work which is difficult is Johnston's *Elements of agricultural chenistry* from the edition of Sir Charles A. Cameron MD, revised and in great part rewritten by C. M. Aikman, Professor of Chemistry, Glasgow Veterinary College (17th edition). The preface is dated December 1893 and the book is one of 482 pp. It was put out by William Blackwood of Edinburgh. It covers the obvious subjects, soils, structure and growth of plants, the improvement of soils, the use of lime, paring and burning . . . animal nutrition, the composition of milk and so on. This is said to have sold 92,000 copies, possibly an optimistic estimate.

I must confess that I am doubtful whether I hve been able to include all that Aikman wrote nor indeed that I have been able to discover all of his rather voluminous output. I can only hope I have done so: I have certainly made the attempt.

Charles Finch Dowsett wrote, or rather edited, *Land, its attractions and riches* by 57 writers, an 8vo of 910 pp. issued by the Land Roll Office, London 1892. It had a slogan "Buy English Acres". It was put out by the author, or rather undertaker, Mr C. F. Dowsett of Winkelbury, Basingstoke, Hampshire, net price £3.6s.5d, post free, being £3.5s for the book plus postage and 1s.5d packing. Pages 1-55, he wrote, contain the contents of the first edition. The contents of the second edition commences on p. 56, chapter XV to chapter XXVII to pages 224, including index. Chapter XXIV, Free trade a failure from the first, by T. Fenn Gaskell. Chapter XXV, Extracts from *Land, its Attractions and Riches* by various writers. Chapter XXVIII, "Warren Hastings, An example of the love of English Acres". Chapter XV, American Land Buyers.

I must confess that I am at some loss to comment upon this voluminous and serious encyclopaedic work, the mere list of contents being quite exhaustive (not to say exhausting!). I shall leave it to the readers, if any, to make their own criticisms and draw their own conclusions about its value to-day and its use when published.

Many of our Victorian ancestors were extremely prolific not only in their intellectual output as well as in the number of their families. I do not know anything about Robert Scott Burn except as a most prolific writer of this and that about farming. I am quite uncertain whether I have captured the details of his numerous publications. I choose to make the first mention of *Systematic small farming or The lessons of my farm; a book for amateur agriculturists*, an 8vo of no less than 334 pp. It was illustrated, and published by Crosby Lockwood. This went into several editions so far as I can ascertain.

Burn also contributed *Outlines of farm management and the organisation of farn labour* to Weale's Scientific and Technical Series of which it is number 207, an illustrated 8vo of 272 pp. put out by Crosby Lockwood c. 1885. This went into several editions. Pt. I dealt with "Soils, manures and crops". Pt. II, Notes historical and practical on farming and farming economy. Pt. III, Stock, sheep, cattle and horses. Pt.4, The dairy pigs and poultry with notes on the diseases of stock. Pt. V, Utilisation of town sewage, irrigation, reclamation of waste land. In the preface to Part II one reads "As adding to the interest of the work it has been deemed advisable to add a few remarks on Roman and Continental agriculture."

In his introduction Burn noted that "It has been said by one of the most graphic writers that agriculture has no history, other occupations are written about but agriculture is left almost entirely to itself": but his current time was a new era (c. 1863). Scientists are working upon it and farmers now want this knowledge. Steam power was being used. Agriculture indeed was in a transition period. All knowledge was less hid in books (though what he means I am at a loss to account for!).

Perhaps the most important work that Scott Burn published was the foolscap work *The Practical Directory for the Improvement of Landed Property, Rural and Suburban and the economical cultivation of its farms adorned with 77 plates and numerous woodcuts*, issued by William Paterson in 1881. This was a quite remarkable work apart from its size.

The preface sounds like a preliminary lecture on the state of agriculture and the adaptations necessary to meet the needs of the time, a new and complete demand with its concomitant results in importation, thus vast lands, dense population (30 million). Their food demands must be met by a wider range of crops in our own fields. He remarks upon the extent of waste and with the contemporary methods unproductive land, which is an assumption that cannot be supported: land reclamation was possible.

The difficulty with a book of this size is that it is almost impossible to read it — like all such magnificent works. To criticise it is even more difficult and trying, but its author possessed the encyclopaedic knowledge (or knew where to find the references) necessary to write it. One thinks of the mere physical effort before the days of typewriters and other machinery including efficient secretaries who could do much of the work; but as a former lecturer he must have had prodigious notes (as I have). The opening chapter is magnificent in its promise of a work that was to be conclusive, as it no doubt was in that day. It opens with a profound but perspicacious declaration i.e. that man and women must eat to live, so that farming is the most important of all industries; everything depends upon it. My comment is the obvious: that without food people starve as indeed so many did in those days of a sort of feudal class structure. Farming, Burn suggested, wanted science, machine skill or experience and all other aids which the modern world would be the last to condemn. The introduction is almost lyrical, and indeed apologetic. Burn remarked that it is not always easy at the outset of a work to detail fully and precisely the detailed subjects which will be treated in its pages. Still, as the work progresses new ideas are elicited, new suggestions made: this proposes to give synopsis an almost impossible task (this chapter is named "Detailed statement of the subject treated"), almost encyclopaedic. He called it a comparatively brief statement of the various divisions of the work. Burn makes so many pertinent remarks that quotation could be too elaborate and indeed too lengthy.

Before this time Scott Burn had gone to Europe with government blessing (I do not know about funds!) and had returned to publish *Notes on an agricultural tour in Belgium, Holland and the Rhine. With practical notes on the peculiarities of Flemish husbandry*, an illustrated 8vo of 241 pp. issued by Longmans in 1862. It would be difficult and perhaps exaggerated industry to write an exegesis on this production and I do not propose to make the attempt.

Burn wrote other rather exotic books which I do not propose to list but I must at least mention *The lessons of my farm. A book for amateur agriculturists,* issued by Crosby Lockwood so long before as 1862 an illustrated 8vo of 334 pp., later issued according to Perkins as *Systematic small farming, or The lessons of my farm. Being an introduction to modern farm practice for small farmers*, Crosby Lockwood 1886, 386 pp. with a frontispiece and other illustrations.

Burn also wrote on architecture, a man of admirable and meticulous industry.

J. Cheal contributed *Practical Fruit Culture. A treatise on practical fruit growing* to Bell's Agricultural Series. It was a small 8vo provided with a frontispiece and other illustrations. It also discussed the tractor (a steam driven machine) especially Proctor's steam digger. It is dated 1892. It must have been a useful little book for fruit growers all over the country but there is of course no means of judging what its circulation was. Its contents are practical, simple and easily intelligible.

Chemical manuring was a burning topic of this time for writers on farming topics. Bernard S. Dyer entered these lists with *How to use nitrate. Practical hints for the profitable application of nitrate of soda as a fertiliser derived from recognised authorities to which is prefaced by permission of the author, "Some Points in Artificial Manuring, A lecture delivered at Lincoln"*, an 8vo of 64 pp., put out by Street and Co. of London. Dyer remarked (p. 16) that Farmers and Gardeners sometimes express surprise that agricultural chemists talk and write so much about nitrogen. When it is known that this is the only element of direct value as plant food it can readily be seen it is entitled to even more attention. The quantity of nitrogen naturally in soils is meagre. Maximum crops to be obtained for any length of time need direct or indirect application of it. The reason farmers do not use nitrate is not because it is not profitable or uncertain but because its merits and regular use are imperfectly understood.

Sufficient has been said to show that this seems a comprehensive work and obviously dealt with a subject that farmers should have been encouraged to study.

Southampton (the Perkins Library) has two books that are not to be found elsewhere, at least by us. They are: W. Sutherland, *Sheep farming, a treatise on sheep their management and diseases*, an 8vo of 163 pp. with plates, issued by W. Cooper, Berkhampsted; and George Fletcher, *The Principles of agricultural practice*, published by the Central Education Co. of Derby, an 8vo of 102 pp., dated 1892.

The social condition of the farm worker (which I have discussed in two books, *From Tolpuddle to TUC,* Windsor Press 1948, and *The English Rural Labourer . . . from Tudor to Victorian Times*, Batchworth Press 1949) exercised a good many pens wielded by thoughtful men. Peter Anderson Graham was one of there. His preface is instructive. "If I were to write a list of the authorities consulted for the making of this volume it would not consist of the names of books and of writers but of obscure men who in the rural districts of England are doing their work and living their lives . . ." He was impressed by the current migration to towns. "Hardly has one family (gone) without knowing why, when and to whither it went . . . As I have wandered in most of the English Shires" (talking to everybody of all classes, or so he said, a rather large claim). The title of his book was *The rural exodus; the problem of the village and the town,* issued by Methuen 1892, an 8vo of 216 pp. It was one of a series edited by H. de B. Gibbins, but this book is social rather than agricultural. Graham wrote another book *The revised English agriculture*, an 8vo of 276 pp. published by Jarrold, London in 1899, but this came out so close to my self-imposed date of 1900 that I have not pursued every enquiry into it.

Every one of the numerous aspirants to fame and to teaching farmers their business was not exactly like Arthur Pillans Laurie who put about his ideas in *The food of plants: an introduction to agricultural chemistry*, an illustrated 8vo of 77 pp. published by Macmillan of London and New York. Laurie was a trifle caustic in his preface in which he wrote that he had "long been convinced that science can be taught only in the laboratory or field. Its value rests only in so far as it presents a logical course of reasoning based on experiment." So this book is written as an experimental introduction to agricultural chemistry for beginners, assuming no knowledge of chemistry. It is for the student himself to perform the experiments under the guidance of a teacher. The book consists of seven chapters each dealing concisely of each facet of food of plants. It was provided with plain diagrams to help the student who can follow the text more easily (an admirable sentiment!)

A list of the apparatus and chemicals required follow. This is considerable and was no doubt costly and it is clear that the guidance of a good teacher would be necessary. The book is quite comprehensive.

I am a litle uncertain whether Herd Books should be included in this bibliography but I am including Grant C. Macpherson's *A history of the Ballindallock Herd of Aberdeen Angus cattle 1861-1891*, an 8vo of 51 pp. issued for Banff at the Banffshire Journal Office 1892. A work of this kind needs no exegesis or explanation.

A book which we could not find in the British Library Catalogue is *Soils and Manures; with chapters on drainage and land improvement* by John M. H. Munro, which was one of a series, Cassell's Agricultural Textbooks, of which the general editor was J. Wrightson. The consequence is we have not seen it!

William Cooper and his nephew wrote *The World's sheep farming for fifty years 1843–1893. In commemoration of the Jubilee of Cooper's sheep dipping powder*, pp. 104. The author claims the survey is the first of its kind. "It makes an encomium on the times". The past 50 years comprise a period which surpasses in interest and activity any other period within the range of history. The book is copiousy illustrated with pictures of sheep, and depicts a Shepherd's Home in Surrey, a hut on wheels. The firm had a 500 acre farm with a celebrated flock of Shropshires. There was some interesting illustrations and a brief exegesis on Russian sheep farming.

Robert Wallace, FLS, FRSE, was the author of *Farm Live Stock of Great Britain*, an 8vo, the 3rd edition of which was published by Crosby Lockwood and Son in 1893. Previous editions had been published in 1885 and 1889, both by Oliver and Boyd of Edinburgh. The preface of the third edition is worth quoting. It runs: "History repeats itself. After the lapse of more than 200 years we find the same spirit of approach to agriculture so quaintly described . . . in 1681. Many men who have received a liberal education deny Agriculture is an Art or if it is of such inferior order that it must not be the same level as other arts".

Again "this new edition is much enlarged and improved and will be found useful by those engaged in teaching under the County Councils." The book opens with a chapter on breeding, crossing and inbreeding. At this time when white in cows was very much disliked it is interesting to note that the author states that a black Galloway and a white bull almost always throw a dark or blue grey.

The writer deals with breeds which are well illustrated. It includes a large amount of detail on pigs, horses, the ass and the mule. Sheep are discussed at length and good illustrations are provided. This is a large heavy volume, useful to isolate the animal and demanding the most careful study by the reader.

John Watson was an editor rather than an original writer. For example in 1892 S. Low of London put out *A handbook for farmers and small holders,* a co-operative work by James Long and other experts which was edited by J. Watson. It was an 8vo of 227 pp., the contents of which are somewhat obvious to the overworked bibliographer. It is always difficult to say which is the most important of the works of such a prolific editor as Watson, but perhaps it is *Ornithology in relation to agriculture and horticulture,* an 8vo by various writers which W. H. Allen of London published in 1893. The book is comprised of eleven chapters each on different birds (species!). Chapter 1 deals with falcons and hawks. The interminable warfare waged by farmers makes one wonder if any will remain (he said) to soar above the fields. They keep mice and other vermin down which are the authors of great depredations, yet the farmers kill these birds. Owls also keep down small rodents, catching more mice than any cat, yet farmers shoot them.

Each chapter deals with specific types of birds. Most take a toll of the farmer's sowing. "The worst to me," he wrote, "is wood pigeon." The good it does by eating grass seeds is outweighed, it seems, by the good seed it enjoys. The book is written by authorities on the birds and makes interesting reading to anyone interested in any of the birds.

Watson also edited one of Rider's Technical Handbooks, i.e. No. 5 of that series, an 85 pp. illustrated 8vo with the title *Farm vermin helpful and hurtful,* by various writers, 1894, which it would be useless to expand upon, the title being so explicit. W. Thacker of London added *The best breeds of British Livestock; a practical guide for farmers and owners of livestock in England and the Colonies,* some 130 pp. 8vo supplied with a frontispiece. The publishers had a branch in Calcutta. He edited other things which are not dealt with here. *A year in the fields*, 8vo pp. 118, had appeared in 1888. It was a normal example of its genre.

Bernard Shirley Dyer, D.SC, FJC, was a successful writer and the author of a 66 pp. 8vo pamphlet entitled *How to use Nitrate. Some points in artificial manuring. Practical hints for the profitable application of Nitrate of Soda as a fertiliser.* It had the unusual addition of an index. To it was prefixed a lecture which had been delivered by Dyer at Lincoln. The pamplet was issued by the Permanent Nitrate Committee, 8 Gracechurch Street, London, EC. It noted that the Committee did not itself sell nitrate of soda and they desire to avoid interfering with the interests of local dealers in fertilizers. Nevertheless it would be happy to give information as to wholesale and retail firms and sources of supply.

A year or two later, the date is imprecisely noted, he published *Fertilisers and Feeding Stuffs: their properties and uses. A Handbook for the practical farmer*, 1893, though even this seems to have been a second edition, a point on which I am not very clear. Another edition seems to have come out in 1894, a 122 pp. 8vo, but I am a trifle uncertain about this. A 3rd edition came out in 1898 and others in 1903, 1908, 1910. It was a comprehensive work, its contents being, I, General functions of fertilisers; II, Farmyard Manure; III, Artificial fertilisers; IV, Applications of Artificial Fertilisers; V, Purchasing Feeding Stuffs; VI, Comparative value of sand. There is an Appendix with notes: I, The Act with notes; II, Regulations of the Board of Agriculture; III, Form of Certificate, and later an index.

One James Sinclair (how many people have borne that name?) edited what might be called sporting books (of which there are many) to which many well-known men contributed, but these are not very much concerned with farming. The first of these (to me) is *Light horses, breeds and management*, one of a series of Livestock Handbooks brought out in 1894 by Vinton, London. It was a 226 pp. illustrated 8vo. It had a frontispiece of a then famous racehorse, St. Simon, engraved by R. Pratt. The book contains pictures of every type of horse, even to the Shetland pony, and the Cleveland Bay originally The Chapman or pack horse, a breed which flourished exceedingly "when the roads in the more remote parts of the kingdom were better". The writers included amongst others William Scarth Dixon, George Fleming and Vera Shaw. Chapter 1 indulges in some history, speaking of Julius Caesar's admiration of some English horses. Some German horses were said, he asserts, to have been given by Hugh Capet to his prospective bridge Elhelevithan, the sister of Athelstan. There is a good deal about imported horses but in general these were not for use on the farms. The book ends with descriptions of some treatment of equine diseases and injuries.

B. Law also corrected a book on *The noble science*, originally written by F. P. D. Radcliffe, but this is outside the present scheme.

Masters in the then growing number of agricultural colleges must have been tempted to write textbooks that they could teach from and so create a market in their pupils. Henry John Webb was one such. He had been Principal of Alpatrai Agricultural College. In 1894 he wrote one of Longmans' series, *An advanced Agriculture*, an illustrated 8vo of no less than 672 pp., 1894. He also wrote *Elementary agriculture; a textbook specially adapted to the requirements of the Science and Arts Dept.* This was one of Longman's elementary science manuals only the 3rd edition of which I have seen. It was published in 1896, an illustrated 8vo of 196 pp. Webb himself announced that, though primarily intended for the Advanced Stage of the Science and Arts Department's Examination in Principles of Agriculture, the book will also cover the greater part of the syllabus for the Honours stage and other competitions such as the Highland and Agricultural Society, the Royal Agricultural Society and so no. It was not apparently intended for practising farmers though it might have been useful to them. The book is divided into two parts: I, Agricultural Science, II, Agricultural Practice. It is comprehensive, the science introduction dealing with such opposed subjects as Geology and Engineering, Chemistry and Botany as well as the anatomy and physiology of farm animals, veterinary science and entomology. In Part II both arable and livestock, dairying and fruit etc. are covered. There are some interesting diagrams and pictures but the work is essentially an examination textbook — perhaps equally valuable to the farmer. The subject is confused between what the farmer wants and what the student wants. It deals with the then up to the minute subject, the process of steam cultivation. It is obviously written by a well informed author, as such a thing should be. The illustrations are interesting.

Here I must turn to what I think is perhaps less interesting. Everybody, or if not everybody most people, know of and are acquainted with Mrs Beeton's cookery book: fewer are aware that her husband Samuel Orchard Beeton cast a gage to fame by writing for farmers. His books I do not propose to analyse minutely but merely to set them out. He started publishing so long before this as 1881. I must confess to some doubt whether I have collected all Beeton's titles but here is the attempt. Beeton's *Farmers' Own Book comprising full and practical instruction on . . . the management of livestock in health and disease for sale and breeding purposes*, an illustrated 8vo of 652 pp. issued by Ward Lock and Co. c. 1898. On pp. 142-3 it says it had previously appeared as *Ward Lock's Book of Farm Management* in 1881. The year 1895 (a guess!) also saw the appearance of Beeton's *Field, Farm and Garden, comprising full and practical information on . . . country sports, tree culture . . . the choice and management of a farm*, an illustrated 8vo of 660 pp. put out by Ward Lock about 1895. Here I may perhaps say that I cannot and do not propose to analyse Beeton's books, because I think their contents are obvious from the titles and they are I suspect little more than exhaustive compilations.

Beeton had published other things before this date. About 1881 Ward Lock published *Corn, roots and other crops of the farm. A practical guide to the successful and remunerative cultivation of wheat oats and barley, rye, etc.*, some 134 pp. an illustrated 8vo. It is quite pedestrian, being the commonplace jobs to be done.

At about the same date Ward Lock also put out *The cow. A guide to dairy management and cattle rearing containing all necessary information regarding animal grazing, milk, butter and cheese*, another illustrated 8vo of 136 pp. Perhaps in 1881 the same publisher issued *Sheep and pigs and other livestock. A complete guide to the breeding and rearing of sheep, pigs, goats, asses and mules*, an 8vo of 118 pp. adorned with a frontispiece and other illustrations. The Perkins Library Catalogue asserts that it was identical with the section *Sheep, pigs and goats etc.* in Beeton's *Farmers' Own Book* and *Ward Lock's Book of Farm Management*. This title is expanded with the assertion that it was *A complete cyclopaedia of rural occupations*, some 1370 pp. 8vo, large enough indeed to be such. The Perkins Library Catalogue asserts that it is a combination of two books published over the name of Samuel Orchard Beeton i.e. *Beeton's Own Farm Book* and *Beeton's Farm Field and Garden*. Separate sections were also published anonymously under various titles. Possibly this list (for it is little more!) is completed by *The Book of garden management. A new edition adapted to the requirements of the present time*, some 760 pp. 8vo garnished with coloured and other illustrations. It is not dated. It opens with the history and literature of gardening. There seems to be nothing novel in it. It is, I can only suppose, a compilation like the cookery and other books of this author and his wife, but it could have been a useful work of reference – a sort of encyclopaedia.

There is also *Beeton's New Gardening Book. A popular exposition of the art and science of gardening . . . thoroughly re-written, revised and re-arranged*, an 8vo with 350 illustrations and some 456 pp. Beeton produced at least 14 gardening books of which a detailed bibliography would be improper here. I may perhaps suspect that my references to Beeton are not complete, but I can only apologise to my devoted readers. These books indeed can, I think, all be dismissed as useful compilations and indeed these and other similar productions may well have been of service to what was no doubt a large number of contemporary readers.

It is always difficult to assess the value of such grandiloquent and omniscient works but at least they may have been useful for reference purposes. It is even more difficult to guess who bought them other than libraries and educational institutions.

Alfred Mansell is another difficult author (perhaps we all are!). He wrote *The formation of pure bred flocks and their subsequent management*, an illustrated 8vo of 62 pp. put out by W. Cooper of Berkhampsted. It is a distinctive and particular piece of work, the details of which are unnecessary here. It deals with the selection of rams, a choice that required some decision. This was dated 1895. In the following year he wrote 14 pp. 8vo on what he entitled, as it certainly was, a *Brief history of Shropshire Sheep with notes on the breed*. It was published by Adnitt and Naunton of Shrewsbury. He also kept a *Private flock book*, the only copy of which in the British Library is a set of blank forms that is of no value, or so I suggest, to the historian.

Many of these writers of farming books and pamphlets are difficult to assess, as must already have been made clear. C. M. Aikman and Patrick Wright translated a German book written by W. Fleischman, the English title of which is *The book of the dairy. A manual of the science and practice of dairy work*, a large 8vo of 344 pp., provided with a frontispiece and other illustrations and an introduction by the English editors. It was published by Blackie and Son of London in 1896.

The introduction states that there had been change and progression in dairying during the previous thirty years. Nevertheless there was a comparative neglect of dairy science. Some pages are devoted to descriptions of the dairy breeds of cattle and complained that very little speculation was devoted to the treatment and management of milk and the manufacture of butter and cheese. Dairying indeed was an art. The book is an exegesis upon the importance of milk production and of butter and cheese. Fleischman protests the importance of butter and cheese. The introduction discusses world trade in milk products. A table of imports of butter, margarine and cheese for 1892-4 is given.

The assertion is that the science of dairying as it is now known had till very recently no existence. Thirty years ago there was practically no dairy literature and no dairy schools.

There was then a dairy literature largely drawn from American and indirectly from German sources. He explained the great advantage of modern and more suitable utensils. There was a large number of new illustrations in the book.

The actual text is quite unexceptional and deals with every branch of the subject of selling milk, cream, butter and cheese. It discusses all the chemical questions involved and there are illustrations, pictures of the apparatus required. At some point is introduced the subject of sheeps' milk and cheese as well as the same produce from goats.

When speaking of the different kinds of cheese, Aikman refers to Pliny and Columella. He supplies a list of comtemporary cheese making in Europe and gives details of their manufacture. The book is completed with a number of tables likely to be of use to readers.

Omnium gatherum works were a fashion in Victorian England. One such was edited by J. Watson in 1892. It was *A handbook for farmers. A collection of articles by experts*, some 227 pp. 8vo. The parts were written by then well known people. "Arable land" was written by James Long, whose son was for many years editor of the Journal of the Ministry of Agriculture. The father was no optimist. Good times had departed, he said on p. 3, and there was then a general move from arable to grass, as indeed is recognised by modern historians, the idea being that expenses were less, and risks not so imminent. Again "The question of what pays, only the farmer can answer for himself". Later stress is laid on the obvious truism that what suits one sort of land is different from that is proper for another.

James Long also wrote *The dairy . . . horse breeding, pig keeping – poultry – fruit culture;* vegetable culture . . . plus tomatoes and vines are discussed and the necessary ways to do them. The section on rabbit warrens for profit fascinated at least one reader. The expense of constructing them as laid down must have been considerable but the way rabbits multiply could perhaps have made this work worthwhile, even profitable. It must have been another useful book to have in that day. It was No. 7 of *Morton's Handbooks of the Farm,* an 8vo of some 148 pages and was illustrated, indeed John Chalmers Morton collaborated in its production. By 1892 it had reached a third edition. It was rather an *omnium gatherum* production.

One G. S. Mitchell wrote *A handbook of land drainage,* an 8vo of some 118 pp., dated 1894. Like so many other contemporary productions it was principally based upon the requirements of the syllabus issued by the Surveyors' Institution. A second edition was put out in 1898, a third in 1910, a statement that perhaps need not be expanded. On pp. 109-115 there is a bibliography of land drainage, a useful guide to the history of the process. Mitchell also wrote *Notes on the valuation of agricultural tenant right intended for the use of students,* a London 8vo of 1900 which need not perhaps be expanded upon at this place.

One James Muir was Professor of Agriculture at Yorkshire College Leeds in 1893 and of course any one holding such a past may be expected and indeed must produce a textbook or so. This man exceeded this requirement by writing *Agriculture, practical and scientific,* an illustrated 8vo of 343 pp. which was published by Macmillan of London and New York. I have described so many books with this sort of title that I think it unnecessary here to expand upon the contents of this one. They are perhaps more or less obvious. Muir also wrote a *Manual of dairy work,* a small 8vo of 93 pp. also published by Macmillan and, finally (for him) in 1899 a pamphlet of 68 pp. on *The utility of sulphate of ammonia in agriculture,* which was promoted by the Sulphate of Ammonia Committee of London in 1899.

This time was one when a good deal of land was being laid down to grass or possibly in some areas being let fall down to grass. William Hutchinson's *Handbook of grasses; treating of their structure, classification, geographical distribution and uses; also describing the British species and their habits,* must therefore have been an acceptable and useful work. It was brief, some 92 pp. 8vo with a frontispiece and other illustrations, which was put out in 1895 by Swan Sonnerenschein in London and Macmillan in New York. In his preface Hutchinson pronounced: "Farmers and graziers to whom cereal and forage crops are all in all rarely have any scientific acquaintance with grasses. This book is an endeavour popularizing their study and importance." It is well illustrated, with Latin names plus the English one, a great help to those knowing Latin. This book would be a great necessity to a student on the subject. It seems comprehensive and is small enough to fit in a bag or packet (where perhaps it was left permanently in peace).

There is necessarily a great similarity between the various writers on such subjects as the dairy. The subject itself changed somewhat but not to a remarkable extent, except perhaps in steam ploughing and gathering the harvest. The introduction of the steam plough, the reefer and binder, and so on (described in my book *The Farmers' Tools*) leaves little room for original comment but emphasises the necessity of repetition by reason of the repetitive character of the operations that had been the same for centuries, as the remaining pictures, in calendars and so on, demonstrate.

The dairy industry was flourishing with the growing demand from the increasing aggregation of human being in the urban sprawl that was an economic condition of the late 19th century. One consequence may have been the output of books of instruction for the aspiring dairy farmer. One of these was an 8vo pamphlet of 80 pp. issued by Bumrose and Sons in 1892. Its title was *Cheese and Butter Making: an elementary course of instruction for technical classes*. Parts I and II of this were written by John Oliver: Part III by Margaret Barrow.

Oliver followed it with *Milk cheese and butter: a practical handbook on their properties and the processes of their production. Including a chapter on cream and the methods of its separation from milk*. This is a substantial volume of 362 pp. 8vo with a frontispiece and other illustrations. It was published by Crosby Lockwood, London.

Oliver wrote in the preface: "Ill heath and the pressure of duty has prevented this work being finished. However the author is happy; after fifteen years of endeavour, it is done." The need for such a volume was constantly felt by other teachers also, and the need for a textbook on dairying was felt by students wanting to master the subject. The book is described as a Handbook for the Dairyer, throughout distinguished from the "Dairy Farmer" or the

"Dairyman"; neither is necessarily a maker of cheese and butter. American and Continental writers strive after attaining the best, but it is necessary for this to be known for the British dairyer to compete. This, he wrote in his first chapter, is the basis for the book, the following six chapters relating to milk, the conditions of its production and its sale. The balance of the book is written on the assumption that these had been carfully studied. The book is comprehensive and to a layman definitely comprehensive. It contains no less than 218 illustrations, a definite help to those finding such a necessity. The author states that he took fifteen years to finally get the work finished and published. It is this concise and tried book that was so valuable to the student — if he read it!

A rather odd production was published by Swan Sonnenschein in 1895. It was a book in the French language entitled (in England ?) *Thorough cultivation; a manual of deep land culture as described by Henry Stephens, Sir Arthur Cotton and others.* It was a 250 pp. illustrated 8vo, but its language perhaps puts it outside the scheme of this bibliography, which, as its title claims, is about *Old English Farming Books.*

I have already said that the farming books of this (and indeed any other) age were necessarily repetitive and that consequently it would (perhaps) be monotonous to give minute detail of the contents of each one. R. Hedger Wallace wrote *White Cattle; an inquiry into their origin and history* which was printed in the *Transactions of the Natural History Society of Glasgow*, new scrics, No. 5, 1897-9, an 8vo illustrated with plates. These cattle had already been the subject of an earlier work, and are a curiosity and rarity, for white animals were disliked by most farmers. Wallace, who lived until the 1930s also wrote a textbook entitled *Agriculture,* an illustrated 8vo of 352 pp. published by W. and R. Chambers of London and Edinburgh. It is a comprehensive work dealing extensively with the different branches of the subject, as would be expected of a writer of the calibre of Wallace. Need I say more!

I shall be more expansive about an Irish born work by Thomas Carroll of Glasnevin who wrote an *Introduction to practical farming; an elementary textbook for use in Irish national schools,* a new edition of which, a 16mo, was printed by Thomas and Co., Dublin in 1896.

Carroll wrote in his preface that the revision of the introduction to practical farming has been done with the idea that pupils of the Irish National Schools could give attention to the principles that underline the various processes of agriculture, and also to help the teachers to give more effective instruction to their pupils. Part I deals with the basic start to agriculture — cultivating the land and the life and development of the seeds planted, the crops to cultivate and their needs. Pt. II deals with the feeding of the land, gardens, flowers, fruit. Parts III, IV and V cover the animals, implements, poultry, dairy, grasses, fences, etc. Although designed for the Irish National Schools the small book would be useful outside that area in our opinion. It is produced in clear type, easy to read script, the sentences numbered and separated so it is made easy to re-read any part. The animals are nicely illustrated, and our criticism is that no unnecessary words are used contrary to the verbosity of so many contemporary farming textbooks and such like.

Alexander Gillespie wrote *The new testament of agriculture*, an illustrated 4to of 272 pp. put out by J. M. Carr, Newcastle upon Tyne in 1895. It was a rather odd production though the major part of its contents followed what had by then become orthodox. The pagination is very irregular, and the last eight, unnumbered pages contain copies of letters dated between 1895 and oddly 1897, a later date than that of the publication itself. It does not, I think, require any futher exposition here.

A botanical work which ought perhaps to have a mention in this compilation was written by one Edward Hackel, and translated by F. Lanson Sinbner and Effie A. Southworth, entitled *The True Grasses*. It was an 8vo of 228 pp. adorned with a frontispiece and other illustrations. It was published by Constable and Co. 1896. No more be said about for it was not only foreign but not strictly speaking "farming".

Primrose McConnell was the author of many books but that of interest here is *The elements of farming,* one of the series *Handbooks of the Farm*, and was intended to be an introductory volume to that series. I have only seen the second edition of this, dated 1896. This author wrote many other books. It was introduced in the preface by the protest "Here the writer has endeavoured to pass in review in as succinct a manner as possible the various departments of farming and farm management". He worked over 600 acres of clay. He knew what he was talking about. Farming ought to be looked at as a broad subject capable of being studied without too much science pure and simple. "An attempt has been made to keep in mind that there is the everyday life of the farmer to be lived under all sorts of conditions and drawbacks", a nice thought, and one that certainly matters.

Soils, crops and livestock are dealt with, accompanied by suitable illustrations in as concise a way as possible. The paragraphs are nicely headed and so easy to backtrack, or be used as reference for the student. Again, as in many other farming textbooks, the hard life of a farmer is highlighted. A modern edition of this book is, I believe, still used by the agricultural community (1985)

This could hardly be true of John Milne and Sons' production *The British Farmers' Plant Portfolios. Specimens of the principal British grasses, forage plants and weeds with full descriptions.* I have only seen the third edition, which consists of 56 pp. of mounted specimens, a folio put out by Simpkin Marshall, London. It is a most beautifully mounted collection of pasture plants. At the time it was published there was more than 20 million acres of permanent and rotation grass lands in Britain. Converting tillage to stock land was continuing. The study of forage plants was not, they said, therefore likely to diminish, but rather to advance. A full acquaintance with their life history and the ability to identify the plants may well be deemed indispensible. Much could be done to improve British pastures, making them more fit to carry more stock. Seed sown pastures were best. Could more be said?

In 1897 Richard Halliburton Adie and Thomas Barlow Wood produced a 2 vol. 8vo with the title *Agricultural Chemistry*, Vol. I of 280 pp. and Vol. II of 229 pp., published by Kegan Paul. I sometimes forget because of advanced years my work and memories of the Research Branch of the Ministy of Agriculture, and fail to realise that books and writers who were familiar to me in the period between the wars are not so to more recent readers. These collaborators were men I knew. Later they wrote *Chemistry for agricultural students*, a book published in 1926 far outside the plan of the present volume of bibliography. In the preface to the earlier book the authors wrote, "the authors like so may others have felt the want of a textbook in teaching the elements of chemistry to the students beginning the study of agricultural science. They hope the book will be found useful by some who have to begin on their own and who would find the larger works on the subject more difficult to handle".

The introductory matter in Chapter I professed "In this study it is necessary to deal with the changes accompanying the growth of plants from the soil and the growth of animals fed with the products of plant life. Firstly to find out the relation of the soil to the wants of the plant, and of that to the requirements of the animal" (a chain reaction). For success each must tie in with the other.

In the frontispiece to Vol. I chemical apparatus is illustrated. These two volumes are very detailed. Paragraphs are carefully annotated in the margin, so looking back for one is easy.

Cadwallader John Bates is a name that ought to be, if it is not, known to all breeders of Shorthorn cattle. He figures in *The history of improved Shorthorn or Durham cattle and of the Kirklevington herd from the notes of the late Thomas Bates. With a memoir by Thomas Bell*, an illustrated 8vo issued by R. Redpath, Newcastle on Tyne in 1871. The same publishers also issued as many years later as 1897 *Thomas Bates and the Kirklevington Shorthorns; a contribution to the history of pure Durham Cattle*, adorned with a frontispiece and other illustrations, an 8vo. In his preface the author states that he is editing the papers and correspondence of his great uncle. Bates himself had intended to write a History of Shorthorns and publish in America. He did not, fearful that his enemies would interpret it as a wish to blaze the fame of the cattle. The book was, he said, written by his tenant Thomas Bell in 1846, and represents all his actual progress. Thomas Bell in 1871 did publish *The History of the improved Shorthorn cattle from the notes of the late Thomas Bates*, but it was not so arranged as to be a lucid history of the Shorthorns or even of Mr Bates.

This is a very 'heavy' book, useful I would say only to the 'gentry'. His (Bates') letters are very detailed. The lineage of "Duchess" is very impressive. Edward, one of Bates' surviving nephews tried to introduce Shorthorns into Germany, where he lived for thirty years, but the German breeders would not pay the price of really good animals. One had then lately insisted on carrying off a young bull and sent a turkey cock in exchange for it. The book has good illustrations of the cattle and, if a reader enjoyed letters, aside from the cattle information, it would be interesting reading!

In 1906 the Rev Matthew Culley edited the *Letters of Cadwallader John Bates*, which was provided with a portrait, an 8vo of 192 pp., which was published by Titus Wilson of Kendal, of which no more need be said here.

Much the same negation applies to Russell, the 11th Duke of Bedford, whose story of *A great agricultural estate, being the story of the origin and administration of Woburn and Thorney*, an 8vo of 254 pp., was issued by John Murray in 1897. It is of course within the scope of this bibliography but its subject is only agricultural because of the activities of this nobleman. Much of its contests are a discussion of contemporary politics and touches only very gently upon farming processes.

Some writers are modest — a dangerous statement! William Spencer Everett was one of them. His gage to fame was *Practical hints on grasses and grass growing in East Anglia*, edited by Nicholas Everitt, an 8vo of 154 pp. published by Jarrold of London in 1897. The author admits that he is no botanist and has avoided the use of botanical names except where he felt it really necessary to do that. Most of the chapters had already been published in the *Field* newspaper, which had given its kind permission to reproduce them. He gives the warmest thanks to correspondents from whom he learned much and to several years of personal observation. For a grass grower to be successful, weeds pernicious to the pastures and the various seeds must be known. All are not dealt with but the selected few are generally a nuisance to the agriculturist. I cannot be too exhaustive about this book although it is a temptation to do so. I can say that it gives good descriptions of grasses and the conditions they need, ending with a note from one correspondent that he considered nettles sometimes make good nurses to young grasses in a draughty season.

Walter Hood Fitch and W. G. Smith produced a book which should no doubt be included here. It was *Illustrations of the British flora; a series of wood engravings with dissections of British plants forming an illustrated companion to the Benthams Handbook of British flora*, only the 4th edition of which I have seen but that is not particularly important because the book is not about farming. To be fully effective the illustrations should have been in colour. No more need be said of this book here. It went through several editions until 1924.

A book bearing the title *The Rothamsted Experiments and their practical lessons for farmers* should need no discussion here because readers of my thesis will be aware of and possibly knowledgeable about the work done at that famous institution. It was an 8vo dated London 1897. It was compiled by C. J. R. Tiffer.

A book that we could not find in the British Library is listed in the R.A.S.E. Catalogue. Its title is *The foreigners in the farmyard*, an 8vo of 1897 but this is not available to us, so I am unable to make any comment upon it.

R. D. Garratt wrote *Practical pig keeping: a manual for amateurs* a 96 pp. 8vo, the second edition of which, revised and enlarged, was published by L. W. Gill, London in 1897. The first had been put out in 1892, a third the same size was published at the relatively late date of 1910. The preface is informative. From 1890 the value of breeding stock and store pigs deteriorated to such an extent that it had been quite hard to find customers at anything like a paying price. Those who sold did so at a sacrifice and those who held on longer to a large stock hoping for better prices far worse. Advancing prices for feeding stuffs meant fewer buyers. The 18 months to the end of 1891 were therefore most discouraging to breeders. In early 1892 the price of feed began to get reasonable. The forced selling of 1891 meant short supply of breeding stock but now (May) there is a better outlook but no statistics are given for pig farmers forced to give up in 1891. There is a brief dictionary of 'pig' terms. Garrett declared that he was not a pedigree pig farmer, but one who looked at pig breeding from the profit angle.

This book certainly seems down to earth and gives precise details. It was probably extremely useful to those into whose hands it fell.

Pigs were always important, from the unit kept by the cottager in a home-made pig-sty to the large herd maintained by a specialist breeder of these brutes. Sanders Spencer appreciated this in his *Pigs: breeds and management. With a chapter on diseases of the pig by J. Wortley Asce and a chapter on bacon and ham curing by L. M. Douglas.* Livestock Handbooks, 5, one of a series issued under the general editorship of J. Sinclair, an 8vo of 180 pp. illustrated with plates, issued by Vinton, London. I hardly think its contents need be expanded here for they are rather obvious from the title.

Liberty Hyde Bailey was the editor of a work, *The principles of agriculture: a text book for schools and rural societies*, a 600 pp. illustrated 8vo put out by Macmillan in 1898. In the preface the author rightly points out the difficulty in teaching agriculture. To the scientist, its application is to agricultural chemistry — to the stockman, raising animals. To horticulturists — fruit growing, flower growing or nursery business, and everyone is certain it is a science since the establishment of agricultural colleges and experimental institutions. The fact is (or so he said) agriculture is pursued primarily for the gaining of a livelihood, not for the extension of knowledge. It is a business, not a science. On that side the experimenter is able to help the farmer. On the business side, the farmer must help himself. Facts of science and scientific thought are essential to good farming.

Again the book deals with all aspects of farming, its primary foundation, the soil — its planting and the animals that are kept. Some of it is difficult reading. The chapter on plants and the crops was written in an easier style, being concise and its paragraphs numbered, making its study easy.

One William Cooper and his nephews resident at Berkhampstead was an inventor of a sheep dip which he may be said to have advertised in his book *The sheep breeders' directory. With hints on cross breeding and of the rearing of sheep for export by the Prop. of Cooper's sheep dipping powder*, an illustrated 8vo of 58 pp. put out by the firm in 1898. In the introduction the authors states "We have utilized the information our worldwide connections enabled us to command" (How true!). The sheep dipping powder seems certainly to have penetrated far and wide.

On p. 2 the authors comment "Fifty years ago the country was emerging from the consequences of the Reform Bill in 1832 and all that followed. It is hardly necessary to observe the impossibility of giving trustworthy figures. After 1866 a system of agricultural statistics was established." Earlier estimates were based on insufficient data.

This booklet is absolutely fascinating aside from the main subject matter. Mr Cooper in 1843 must have been blessed among farmers for the Cooper Dipping Powder which went far and wide. Not only is the information good but the illustrations taken from photographs or papers are able to convey the way of life, the size of flocks of the country's smallest colony, the Falkland Islands, which is described as so vague that even the geographical position is probably unknown.

A 34 pp. 4to by one Thomas McKenny Hughes was issued by Nicholas, Westminster in 1896. It was extracted from *Archeologia* no. 55, 1896, pp. 125-158. Its title was *On the more important breeds of cattle which have been recognised in the British Isles in successive periods and their relation to other archaelogical and historical discoveries, as presented to the Society of Antiquaries*. For some reason we could not find this, a bad example perhaps of incompetence.

Jersey cattle have always been the object of admiration, not least to those English farmers who kept them. The English Jersey Cattle Society through Vinton the publisher put out a guide to their owners entitled *Jersey cattle, their feeding and management. Compiled from information received from members of the English Jersey Cattle Society and published for the Society* in November 1898. The contents I cannot discuss because I could not find this production in the British Library and do not think it is available there.

Herman Biddell was one of a group of writers who contributed to a series *Livestock Handbooks* of which the general editor was J. Sinclair. His contribution was no.3 *Heavy Horses; breeds and management* an illustrated 8vo of some 219 pp. He confessed that it was very difficult to give a short history. The Romans viewed the breed as unusually large. No plates or drawings exist of the very early horses but Biddell thought they were of considerable bulk. Their descendants were decidedly so and bore a close resemblance to the Shire Horse of the these days. These had to carry armed men over rough wooded and rugged terain, so size mattered.

Chapter IV of the book was written by Mr W. R. Trotter who said that the breeding of cart horses for street work was one of the most important departments of British farming. At that time − 1893 − when a drought was current and meat prices were low heavy horses still fetched high prices.

Diseases and injuries to which heavy horses were liable are annotated. Emphasis is placed upon the farmer not to take chances but call in a vet's However symptoms are noted and a first aid treatment is suggested that could make a vets' visit unnecessary, but careful assessment was necessary.

This seems a very comprehensive work. Good illustrations emphasise how necessary it was for a farmer to add doctoring and careful nursing, and to care for horses off colour or really ill, to his other attainments.

This was one of a series of *Livestock Handbooks* numbering five in all, being John Wrightson, *Sheep breeds and management*, 1893; W. C. A. Blew and others, *Light horses, breeds and management*; Biddell's book; William Houseman, *Cattle breeds and management with a chapter on diseases by J. Wortley Axe*, 1897; and Sanders Spencer, *Pig breeds and management*, 1897, which ran into five editions, the fifth dated 1900.

Insects, foes and friends were of importance to the farmer. William Fursell Kirby, who exercised his ingenuity and knowledge in writing a book about them, a London 8vo of 1898. The description of the size and number of these is quite frightening, or so he said. In this small volume the author has annotated those injurious to our economy. The coloured illustrations are extremely good and would be of value to those studing the subject. Species too rare to become injurious are included. There seemed no reason why they should not. The Wild Birds Protection Act had he said, reduced the number of many insect species. A note propounded the puzzle "Note the letterpress has been translated and adapted by W. F. Kirby, FLS, FES etc.", but nowhere could we see any reference to another language. Probably insect names were given in Latin and English. A very helpful book, this, with suggestive ways of getting rid of these pests involving much effort and time.

Another book which would probably have been interesting was destroyed by bombing in the war. It was *Estate fences, their choice, construction and cost. And a chapter on boundaries and fences in their legal aspect* by T. W. Marshall, the book itself being written by Arthur Vernon. The records state that it was an illustrated 8vo of 420 pp. published by E. and F. N. Spon, London in 1899.

There were still some urban dairymen at the end of the 19th century. One of these was inspired to write a book. He was William Smith, dairyman of Edinburgh whose book was *A Practical guide to dairying*, an illustrated 8vo of 145 pp. put out by Banks of Edinburgh in 1900. The preface states that the book had been written to meet a continual flood of enquiries from various parts of the world (no less!). The author says he has long dealings with the dairy trade and a comprehensive knowledge of all its branches as well as the machinery employed. He declared that every country and district kept to the breed that suited it. Not every breed was adapted for dairy purposes. Some breeders use three breeds of cattle. The great importance is to use only pure-bred cattle. The characteristics of the cow best suited are pointed out. As to feed, the animal's condition is a guide to the amount to be given. It is interesting to note that in towns where there are brewers, there are often a number of still-fed cows kept on the by-product, brewers grains' and dregs. The average life of a cow under this treatment from the date of calving is eight months.

A warning is issued that building dairies is costly, and they are expensive to equip properly; a great deal of careful study is necessary before embarking on such a venture.

Thomas Southall Dymond wrote *An experimental course of chemistry for agricultural students,* an illustrated 8vo of 192 pp. It was published by E. Arnold of London in 1898 as one of their series *Arnold's Practical Science Manuals,* of which the general editor was

R. Meldola, who added a note to the effect that Education in Science is a difficult subject for the teacher to deal with since so many diverse views are held on the subject, representing the different methods of treatment to which various branches of science lend themselves. A reasonable assumption is that those concerned know something of the general principles of Chemical Science. Elementary schools especially in country districts are feeble. There is a crying need for good secondary schools in rural areas, a need in which no progress has yet been made. Scientific principles should be part of the early training of boys and girls. In most parts of the country where County Councils with funds have placed such instruction within reach, a young generation of farmers have availed themselves of them.

Small drawings help the student to follow the experiments under teacher supervision. This seems to be a useful little book though some of the print is very small for even good eyesight. Questions and problems are set out at the end of the chapters.

A textbook of plant diseases caused by cryptogamic parasites, an illustrated 8vo of 458 pp., was issued by Duckworth of London and MacMillan of New York in 1899. The author also wrote several other books on analogous subjects. Perhaps it will be sufficient here to give their titles, i.e. *European fungus flora*, 1902, *Diseases of cultivated plants and trees*, 1910, *British fungi, with 40 coloured plates of Mildews, Rusts and Smuts*, 1913. I could, of course, be very expansive on this subject but I do not think I ought to indulge myself in this place.

The century ended with the works of two distinguished writers, Sir Walter Gilby, Bart., and Rider Haggard. The first of these was the author of *The Great Horse — The Shire Horse or the War Horse from the time of the Roman invasion until its development into the Shire* only the second edition of which I have seen. This was a 69 pp. illustrated pamphlet issued by Vinton. The preface states that since the publication of the first edition (which I regret to say I have not seen) great progress has been made on the improvement of the Shire Horse so it has been remodelled and enlarged. The horse known as the Shire Horse is the purest survival of the type described by medieval writers as the Great Horse, a development of the Ancient British War Horse admired by Julius Caesar. English men, wrote Sir Walter proudly, have achieved many triumphs as breeders of domestic animals and none more so than the heavy draught horse. The heavy horse has been closely identified with the lot of the people of Britain from earliest times. It must not be forgotten that its use in agriculture is relatively modern. Until the Middle Ages almost all heavy draught work on the farm was performed by oxen. The history of the horse is given in great detail with illustrations. There are interesting newspaper advertisements of horses for sale.

Gilby wrote several other books of perhaps less importance but some were published after 1900 so I am uncertain whether to include them. One was *Farm stock a hundred years ago*, an 8vo of 154 pp. published by Vinton in what I think is slightly outside my chosen period.

Gilby wrote a number of other books, some of which were published after 1900, but deserve at least a note here. One was *The Pig in health and disease. How to avoid swine fever*, a 46 pp. illustrated 8vo put out by Vinton (? 1907).

In 1900 appeared his *Animal Painters of England from the year 1650*, illustrated with wood engravings by G. F. Babbage, a 2 volume production dated 1900. Another was *Horse breeding in England and India and Army Horses abroad*, a London 8vo of 1901.

The Highland and Agricultural Society Catalogue lists *The harness horse*, an illustrated London vol. of 1898; *On the care of horses* London 1898 and *Young Race Horses*, also London 1898, but these I have not seen.

Rider Haggard is more famous, if he is not now forgotten, for his romances than for his farming writings but these are of more importance here. He is a writer who outspans my period but writers of the latter years of the century are apt to linger on into the 20th century. *The Farmers' Year, being his commonplace book for 1898* an illustrated 8vo of 489 pp. was issued by Longmans in London and New York in 1899. It is supplied with 2 maps and 36 illustrations. First printed in September 1899, it was reprinted in the following November and in the Silver Library in 1906. It was never intended to be a manual of farming, but was rather a diary of events rather than a calendar of the processes, a record of one year of the daily experiences and reflections of an individual, but this is not the place to be expansive about his ideas and reflections nor his exegesis upon the history of his holding.

Besides this, Haggard was engaged to make a survey of the condition of farming in this country. His observations are recorded in *Rural England being an account of the agricultural and several researches carried out in 1901-2*. This is supplied with maps and illustrations and was issued by Longmans in two volumes, 8vo, with pp. 584 and pp. 623.

He continued to write and be published into the 20th century but his books of that date may perhaps be omitted here. I am not sure that is justified, so I give the titles. *A gardener's year,* a 404 pp. illustrated volume issued by Longmans in 1905. *The poor and the land, being a report on the Salvation Army colonies in the U.S. and at Hadleigh, England, with a scheme of national land settlement*, an 8vo of 157 pp. issued by Longmans in 1905. *Rural Denmark and its lessons* was published in 1913, but I think I must confine myself to these words.

It has taken a long time to compile and write this production but it was a job that intrigued and indeed fascinated me, and it is, I hope, a worthy successor to the previous volumes I have compiled upon this subject; and so I conclude these studies at the turn of the century, the year 1900.

BIBLIOGRAPHY

(a) *Agricultural Bibliography and Critical Work.*
  i. Haller, *Bibliotheca Botanica*, 1772.
  ii. Richard Weston, *Tracts on Practical Agriculture* . . . with . . . *a Chronological Catalogue of English Authors on Agriculture, Botany, Gardening, etc.*, 2nd ed., 1773.
  iii. Richard Pulteney, *Historical and Biographical Sketches of . . . Botany*, 1790.
  iv. John Lawrence, *A Philosophical and Practical Treatise on Horses*, 1796.
  v. Richard Weston, 'Various gardener's calendars published in England', *Gentleman's Magazine*, Dec. 1804. See also vol. for 1806.
  vi. Idem, 'Critical remarks on botanical writers', *Ibid.*, 1807.
  vii. Samuel Egerton Brydges, *Censura Literaria*, 1806.
  viii. John C. Loudon, *Encyclopaedia of Gardening*, 1822.
  ix. Idem, *Encyclopaedia of Agriculture*, 1825.
  x. Samuel Felton, *On the portraits of English authors on gardening, with bibliographical notices of them*, 1829.
  xi. George William Johnson, *History of English Gardening, Chronological, Biographical, Literary and Critical*, 1829.
  xii. Cuthbert W. Johnson, 'The Early English Agricultural Writers . . .', *Quarterly Journal of Agriculture*, xii et seq.
  xiii. John Donaldson, *Agricultural Biography*, 1854.
  xiv. R. W. Blenceux, 'Notes from old books', *Journal of the Bath and West Society*, N. S. vol. v, 1857 et seq.
  xv. Donald G. Mitchell, *Wet days at Edgewood*, 1884. (Originally published 20 years before.)
  xvi. Benjamin Daydon Jackson, *Guide to the Literature of Botany*, 1881.
  xvii. Albert Forbes Sieveking, *The Praise of Gardens*, 1885.

xviii. W. Carew Hazlitt, *Gleanings in Old Garden Literature*, 1887.
xix. F. H. Huth, *An Index to Works on Horses and Equitation*, 1887.
xx. Earl Cathcart, 'Jethro Tull, his Life, Times and Teaching', *Journal of the Royal Agricultural Society of England*, 3rd ser., Part ii, 1891.
xxi. John D. Sedding, *Garden craft, old and new*, 1891.
xxii. Russell M. Garnier, *History of the English Landed Interest*, vol. ii, 1893.
xxiii. Mrs. C. W. Earle, *Pot Pourri in a Surrey Garden*, 1897. 1st and 3rd vols.
xxiv. A. M. Earle, *Old Time Gardens*, 1902.
xxv. Rose Standish Nichols, *English Pleasure Gardens*, 1902.
xxvi. Patent Office, *Subject List of Works on Agriculture, Rural Economy, and Allied Sciences*, 1905.
xxvii. Donald Mc Donald, *Agricultural Writers 1200–1800*, 1908.
xxviii. Hon. Alicia Amherst (later Hon. Mrs. Evelyn Cecil), *A History of Gardening in England*, 1st ed. 1895, 2nd ed. 1910.
xxix. F. W. Oliver, *Makers of British Botany*, 1913.
xxx. Anon., *Catalogue of the Royal Agricultural Society of England Library*, 1918.
xxxi. Maj. Gen. Sir Frederick Smith, *The Early History of Veterinary Literature and its British Development*, vol. ii, 1924.
xxxii. G. D. Amery, 'The Writings of Arthur Young', *Journal of the Royal Agricultural Society of England*, 1925.
xxxiii. Mary S. Aslin, *Catalogue of the Rothamsted Library*, 1926. See also 2nd ed., 1940.
xxxiv. Royal Horticultural Society, *The Lindley Library*, 1927.
xxv. G. E. Fussell, *Chronological List of Old Books in the Ministry of Agriculture Library*, 1930.
xxxvi. J. A. S. Watson and G. D. Amery, 'Early Scottish Agricultural Writers', *Transactions of the Highland and Agricultural Society*, 5th ser., vol. xliii, 1931.
xxxvii. E. S. Rohde, *The Story of the Garden*, 1932.
xxxviii. E. H. M. Cox, *History of Gardening in Scotland*, 1935.
xxxix. A. McCallum, Various essays in the *Scottish Journal of Agriculture*.

BIBLIOGRAPHY

xl. W. Frank Perkins, *British and Irish Writers on Agriculture*, 3rd ed. 1939.
xli. R. C. Punnett and E. C. Lewer, *Notes on Old Poultry Books*, Feathered World, London 1930.
xlii. F. A. Buttress, *Agricultural Periodicals of the British Isles, 1681–1900*, University of Cambridge, School of Agriculture, 1950.
xliii. The Royal College of Veterinary Surgeons. *Catalogue of the Historical Collection*, 1953.
xliv. The Royal Highland and Agricultural Society. *Library Catalogue*, 1953.
xlv. Southampton University Library. *Catalogue of the Walter Frank Perkins Library*, 1961.
xlvi. F. N. L. Poynter, *A Bibliography of Gervase Markham*. Oxford Bibliographical Society Publications, N. S. Vol. II, 1962.
xlvii. Bath and West and Southern Counties Society. *Catalogue of the Library*, 1964.
xlviii. The Royal Veterinary College. *A Catalogue of the Books, Pamphlets and Periodicals up to 1850 in the Library*, 1965.
xlix. *Historic Books and Manuscripts Concerning General Agriculture in the Collection of the National Agricultural Library*, Washington, D.C., U.S.A., 1967.
l. *A Catalogue of the Bradshaw Collection of Irish Books in the University Library Cambridge*. 3 vols. 1916.
li. *A Catalogue of the Agricultural and Horticultural Books 1543–1918 in Wye College Library*. The Library, 1977.

(b) *General Bibliographies that include Works or Sections on Agriculture.*

i. Watts, *Bibliotheca Britannica*, 1824.
ii. Lowndes, *Bibliographer's Manual of English Literature*, 1824.
iii. Allibone, *Dictionary of English Literature*, 1870.
iv. Halkett and Laing, *Dictionary of Anonymous and Pseudonymous Literature of Great Britain*, 1882.
v. Kirk, Supplement to Allibone, 1891.
vi. Gray, *General Index to Hazlitt's Handbook*, 1892.
vii. Charles A. Stonehill, Jun., Andrew Block and H. Winthrop Stonehill, *Anonyma and Pseudonyma*,

1926.
viii. Judith Blow Williams, *Guide to the Printed Materials for English Social and Economic History*, 1926.

INDEX OF AUTHORS

Acland, Arthur H. D., 17, 18
Agricola, 66
Aikman, Charles M., 104–5, 115–6
Anderson, Thomas, 8
Andrews, George Henry, 5, 6
Archer, Thomas C., 23
Armitage, George, 100–101
Ashburner, Robert W., 91–92

Baker, T. H., 66
Bate, Thomas, 11
Beavor, Rev. William H., 73–75
Beeton, Samuel O., 113–115
Blair, Mrs. Ferguson, 11
Bland, William, 18
Bradough, Charles, 85
Brooks, Rev. George, 75
Brown, Robert Erskine, 32, 33
Brown, Sir George T., 78
Brown, William, 37, 38
Buckmaster, John C., 62
Bunyard, George, 61
Burges, Y. H., 31
Burke, Mick R., 43
Burke, Ralph Ulick, 39
Burn, Robert Scott, 106–8
Burness, William, 78–79
Burton, Katharine, 75–76

Cantrell, Charles L., 31
Cargill, Thomas, 33
Carrington, William T., 76
Cheal, J., 108
Christy, Thomas, 49, 50
Church, Arthur H., 31, 64–65

Clarke, John Algernon, 31
Clements, Hugh, 58–59
Clifford, Frederick, 46
Coleman, John, 46
Concalon, Victor, 6
Constable, John, 19, 30, 31
Cooke, Mordecai C., 79
Cooper, William, 110
Copeland, Samuel, 25, 26
Corfield, William H., 38
Coulton, Miss, 40
Cox, Irwin E., 29
Craig, James, 44
Crews, A. W., 55
Crookes, Sir William, 39, 40, 101–102
Cruikshank, W. W., 102
Curror, David, 45

Day, William, 92, 95
Dean, George Allen, 4
Denton, John Bailey, 24, 25, 31
Dowsett, Charles Finch, 106
Drummond, James, 16
Dryland, Peter, 12, 13
Dyer, Bernard, 108–9, 112
Dyer, W. T., 31

Edmonds, W. J., 31
Elliot, Thomas J., 69
Ewart, John, 4, 7

Falk, Robert, 26
Fisher, John, 22
Fitt, James Neville, 28

Fleischman, W., 115–6
Fletcher, George, 109
Fletcher, Thomas, 13
Fream, William, 98–99
Fry, George, 77
Fyffe, W. Wallace, 8

Garnett, Thomas, 77
Glenny, George, 6
Gould, John P., 45
Graham, Peter A., 109
Grant, James F., 7
Grieve, John, 4, 5
Griffiths, Arthur B., 99–100

Hanham, Frederick, 69
Harris, J. B., 77
Harris, William Henry, 23
Harrison, J. T., 31
Harvey, Alexander, 51
Haywood, James, 4, 5
Heatley, George S., 65
Hill, Arthur J., 31
Hutchinson, William, 117–118

Ingham, Hastings, 85–86

Jackson, Henry Kains, 92
Jamieson, Thomas, 48, 49

Kebbel, Thomas E., 102–3

Laurie, Arthur P., 109, 110
Law, B., 113
Lawson, William, 46, 47
Lewes, Sir John B., 41
Lloyd, F. J., 98
Lock, Alfred G., 26, 27
Long, James, 62, 63, 116, 117

MacDonald, Duncan G., 20, 21
Macdonald, James, 80–81
MacPherson, Grant C., 110
Magne, M. M., 27, 28
Main, James, 2, 3
Malden, Walter J., 103–4
Manning, Prentice, 86

Mansell, Alfred, 115
Martin, W. C., 81
Martineau, Harriet, 14, 15
Masters, Maxwell T., 67, 68
McAlpine, Archibald, 95
McConnell, Primrose, 67
Mirehouse, H. J., 103
Mitchell, George, 47
Mitchell, G. S., 117
Morton, John Chalmers, 63, 78, 96, 97
Muir, James, 117
Munro, John M. H., 110

Nevile, George, 70

Omerod, Eleanor Anne, 93

Peard, William, 32
Pilley, John J., 59
Pink, James, 53, 54
Potter, Thomas, 81
Preston, Samuel, 86
Pringle, R. D., 15
Pringle, R. O., 35, 36, 42
Prout, John, 59, 60

Ramsay, Alexander, 54
Ransome, J. E., 31
Redwood, Theophilus, 28
Reid, William, 42, 43
Rogers, A. G. L., 68
Rogers, Thorold, 68
Roland, Arthur, 50, 57, 58, 87, 88

Scott, John, 82–83
Sellar, William, 42, 43
Sheldon, John P., 56, 57
Sheriff, Patrick, 47, 48
Simmonds, Peter, 51
Sinclair, James, 112
Smith, Alexander, 15
Smith, Edward, 49
Smith, George H., 67
Spooner, Lucius H., 3
Squary, Elias P., 31

INDEX OF AUTHORS

Stables, Alfred, 42
Stanford, Edward, 19, 20
Stephens, Henry, 29
Stevenson, David, 45
Storer, John, 54, 55
Stubbs, Charles W., 70, 71
Sturge, William, 17
Sutherland, W., 109
Sutton, Martin John, 83, 84

Tanner, Henry, 56
Tarland, Dran, 51
Taylor, John Elton, 65
Taylor, William Charles, 77
Thacker, W., 111, 112
Thier, M. A., 38
Toms, Frederick W., 68

Upton, H. M., 93
Usher, John, 48

Valentine, Charles R., 97, 98

Ville, Georges, 39

Walker, John, 84, 85, 90, 91
Wallace, Robert, 110, 111
Walley, Thomas, 91
Ware, Lewis Sharp, 56
Warrington, Robert, 31
Watson, John, 111, 112, 116
Webb, Henry John, 113
Welford, R. G., 31
Wheeler, W. H., 94
Willis, J. J., 62
Wilson, John, 16
Wilson, William, 98
Wood, Theodore, 94, 95
Woods, Henry, 21, 71, 72
Wright, Sir R. P., 72, 73, 115
Wrightson, John, 31, 52

INDEX OF TITLES

The Abbotts Farm, 1881, 56
Advanced Agriculture, 1894, 113
Advice to Youths, N.D., 58
Agricultural Chemistry, 1897, 121
Agricultural Education, 1890, 67
The Agricultural Lockout, 1874, 46
Agricultural Education. What is it? 1864, 18
Agricultural Elementary Course, 1891, 103
Agriculture. Its History, 1887, 85
The Agricultural Labourer, 1870, 102
Agricultural Notes on Hertfordshire, 1864, 12
An Agricultural Notebook, 1885, 77
Agricultural Textbook, 1877, 52
Agricultural Science applied in Practice, 1859, 8
Agricultural Series, 1896, 103
Agricultural Statistics, 1870, 38
Agriculture, Ancient and Modern, 1866, 25
The Agriculture of Berkshire, 1861, 11
Agriculture. A Poem in 16 Parts, 1854, 15
Agriculture in Relation to Chemistry, N.D., 89
Agriculture. Theoretical, 1919 reprint, 52
The Agriculturists Assistant. Principal Rules and Tables, 1857, 7
The Agriculturists Calculator, 1867, 30
Agriculturists Non Superphosphate Makers, 1872, 26
An Alphabetical Arrangement, 1885, 73
Among the Clods, 1884, 71
Animal Food Resources, 1885, 51
Animal Painters, 1900, 129
Annual of Philosophy, 1883, 77
The Apple and Pear, 1886, 80
Arboriculture for Amateurs, 1880, 58
Archaeologia, 1896, 125
Artificial Manures, 1867, 39

Beeton's Farm, Field and Garden, 114
Beeton's New Gardening Book, 115
Beeton's Own Farm, 114
The Best Breeds, 111
The Best Forage Plants, 1889, 95
Black Faced Sheep, 83
Book of the Diary, 1896, 73, 115
Book of the Farm, N.D., 33
Book of Farm Implements, 33
Book of Farm Management, 1881, 60, 114
Book of Garden Management, 114
Book of the Horse, 70
The Book of Landed Estate, 1869, 32
The Book of the Pig, 1886, 62
Book of Points of the Pig, 1870, 23
Book of Rothamsted Experiments, 1905, 72
Border Breeds of Sheep, 1875, 48

INDEX OF TITLES

The Breeding and Management of Pigs, 1865, 22
The Breeding and Management of Sheep, 1864, 21
Brief History of Shropshire, 1896, 115
Brief Notes, 1894, 72
British Dairying, 1893, 57
British Dairy Farming, 1885, 62
British Farming. A Description of the Mixed Husbandry, 1862, 16
British Fungi, 1913, 128
British Rural Sports, 1890, 110
British Sheep Farming, 1870, 37, 41
Buildings for Small Holdings, 1909, 81
Butter Making, 1888, 97

Canadian Agriculture, 1893, 63
Cattle Breeds, 1897, 126
Cattle and Cattle Breeders, 1869, 35
Cattle Food Adulteration, 23
Cattle and their Management, 101
Cattle Management, 1862, 15
Cattle, Sheep and Deer, 1872, 21
The Cereals, 1851, 51
Cheese and Butter, 1892, 118
Cheese and Butter Making, 1896, 63
Chemistry and Crops, 1882, 76
The Chemistry of Agriculture, 1874, 45
The Chemistry of the Farm, 1881, 72
Chois des vaches laitières, 38
Clodhopper Cracks, 1872, 45
A Common Sense View of the Potato Disease, 1872, 44
Complete Cyclopaedia, 114
Complete Grazier, 99
Compulsory Cultivation, 1887, 85
The Construction of Silos, 1886, 81
Contribution to the Land Question, 1883, 50
Conversion of Arable Land, 103
Corn Roots, 1881, 114
The Cottiers Cow, 1880, 85
Country Gentleman's Magazine, 1974, 85
The Cow, 1870, 35
Cow and Calf, 1886, 84
The Cow and Calf, 1888, 90
Cows and the Dairy, 90
Crops of the Farm, 1886, 78
The Cultivation of Land by Steam Power, 1870, 42
Culture of the Apple and Pear, 1797, 80
Cultivation of the Soil, 33
Cumberland Farm Life, 32

The Daily Life of Our Farm, 1870, 74
The Dairy, 38, 117
The Dairy and Butter Making, 1891, 102
The Dairy Farm, 1889, 63
Dairy Farming, 1880, 53
Dairy Farming and Management of Cows, 1879, 57
Dairy Farming Management, 87
Dairy Stock. Its Selection and Diseases, 1861, 14
A Digest of Facts, 1870, 38
Diseases of Crops, 1890, 99
Drainage and Embarking, 1884, 82
Drainage of the Fens, 94
Drainage of Land, 1880, 88

The Education of the Farmer, 1857, 18
Elementary Agriculture, 1896, 113
Elementary Textbooks, 1884, 73
Elements of Agricultural Chemistry, 1860, 1881, 1893, 8, 104, 105

Elements of Agricultural Chemistry, 54
Elements of Agriculture, 1918, 99
Elements of Dairy Farming, 1894, 63
The Element of Farming, 1896, 120
The Elements of Farming, 1902, 67
English Guernsey Cattle, 1885, 78
The English Rural Labourer, 1949, 109
English Trees, 1880, 58
Ensilage, 1883, 50
Ensilage in America, 1883, 68
Ensilage, Instructions, 1877, 68
Ensilage: Its Origin, History and Practice, 1883, 22
The Equipment of the Farm, 1884, 78
Essays in Natural History, 1883, 77
Estate Fences, 1899, 127
Etymological and Pronouncing Dictionary, 1871, 43
European Fungus, 1902, 128
Evil Results of Overfeeding Cattle. A New Inquiry Fully Illustrated by Coloured Engravings, 1858, 7

Fallow and Fodder Crops, 1889, 52
Farm Buildings for Landowners, 1860, 53
Farm Crops, 1891, 52
Farm Field, 1893, 103
The Farm, Garden, Stable and Aviary, 1869-1871, 29
The Farm Homesteads of England, 1865, 24
Farmlife or Sketches for the Country, 1861, 13, 15
Farm Live Stock, 1893, 110
Farm Roads, Fences and Gates, 1883, 82
Farm Stock, 129
Farm Vermin, 1894, 111
The Farmer, 58
The Farmers' Assistant, 1852, 4
The Farmer's Friends, 1888, 94
The Farmer's Harvest Companion, 1870, 79
Farmers Own Book, 1898, 114
Farmers Year, 1898, 129
Farming for Pleasure, 54
Farming for Pleasure, 8 Vol, 1879-1881, 84
Farming to Profit, 1888, 91
Farming in a Small Way, 1881, 62
The Farming of Somersetshire, 1851, 17
Farming World, 82
Farms and Farming, 1884, 70
Farmyard Manure, 1892, 104
Fertilisers, 1893, 112
Field, Farm and Garden, 1895, 114
Field Implements, 1884, 82
The Fields, 1881, 93
The Fields of Great Britain, 1881, 58
Food of Crops, 1895, 105
The Food of Plants, 109
The Food of Plants, a Plain Treatise on Manures, 11
Forage Plants, 1877, 50
Foreigners in the Farmyard, 1897, 123
The Forest Planters and Pruners Assistant, 1847, 3
Formation of Pure Bred, 1895, 115
Fruit Farming for Profit, 1881, 61
Fungi, their Nature, 1886, 79
Future of British Agriculture, 1893, 57

The Great Horse, 128
The Grouse Disease, 1883, 21

INDEX OF TITLES

Guide to the Cultivation of the Potato, 1879, 53
A Guide to Dairy, 1881, 114
Guides to Methods, 1884, 93
A Handbook for Farmers, 1892, 111, 116
Handbook of Grasses, 1895, 117
Handbook of Insects, 1898, 93
A Handbook of Land Drainage, 1894, 117
Handbook of Sewage Utilisation, 1872, 44
Handbooks of the Farm, 1882, 63, 67, 72, 76, 78, 82, 120
The Harness Horse, 1898, 129
Harvesting Crops Independently, 66
From Hay Time to Hopping, 1860, 41
Health, Husbandry and Handicrafts, 1861, 14
Hearth and Homesteads, 1867, 28
Heavy Horses, 126
The Henwife, 1861, 11
Herefordshire Pomona, 1876, 80
High Farming, 1867, 28
Hints on Breeding, 91
Hints for Farmers and Useful Information for Agricultural Students, 1869, 9
Histoire de l'Agriculture, 1857, 6
Historical Sketch, 1884, 71
A History of the Ballindalloch Herd, 1892, 110
The History of the Clydesdale, 1884, 69
A History of the Fens, 1894, 94
History of the Highland & A. S., 1879, 54
History of Polled Aberdeen, 1882, 80
A History of Prices, 1866-1902, 68

My Home Farm, 1883, 75
The Home and Foreign Agricultural Miscellany, 27
Hops, their Cultivation, 1877, 51
Hops and Hop Pickers, 77
The Horse, 1890, 96
Horse Breeding, 1900, 129
The Horse, How to Breed, 92
The Horse and its Rider, 1861, 101
Horses, Sheep, Pigs and Poultry, 1870, 38
How Crops Grow, 1869, 35, 64,
How to Farm with Profit, 1910, 91
How to Know Grasses, 95
How Plants Grow, 65
How to Select Cows, 96
How to use Nitrate, 108, 112
Hunting, Steeplechasing, 1867, 28

Illustrations of Vegetable Physiology, 1833, 2
Improvement of the Cereals, 1873, 47
The Industrial and Commercial History, 1894, 68
The Influence of Phosphates, 1901, 73
Inter Alia of Farming for the Million, 1854, 6
An Introduction to the Chemistry of Farming, 1892, 18
Introduction to Practical Farming, 1896, 119
Investigations into Applied Nature, 1896, 98

Laboratory Guide, 1882, 64
Land Agents Record, 1887, 86
Land, its Attractions, 1892, 106
The Land and the Labourer, 1884, 70
The Land Steward, 1851, 4
Land, Technical Instruction in

Agriculture, 1882, 52
The Land Question, 1884, 69
Lecture on Abortion and Mortality amongst Ewes, 1883, 22
Lectures on Breeding, 1864, 71
A Lecture of the Breeding and Management of Sheep, 1864, 21
A Lecture of the Diseases, 1873, 72
A Lecture on Ensilage, 1884, 22
Lessons of my Farm, 1862, 108
A Letter to W. Miles Esq., M.P., 1850, 17
Letters to Cadwallader, 1906, 122
Letters to Farmers, 1852, 5
Light Horses, 1894, 112
Life on the Farm, Plant Life, 1883, 67
Livestock, 1892, 52
The Livestock of the Farm, 1874, 36, 76
Livestock Handbooks, 1893, 126
Livestock in Health, 1902, 57
Livestock Journal, 1886, 35

Manufacture of Beet Root Sugar, 1870, 39
The Management of the Dairy, Pigs and Poultry, 1863, 10
Manual of Dairy Work, 1899, 117
Management of Fattening Cattle, 1870, 38
Management of Grassland, 1881, 88
The Management of a Home Farm, 1869, 78
Manures, 1880, 55
Manures and their Uses, 1889, 99
Market Garden Husbandry, 1881, 88
Meat, Milk and Wheat, 1857, 18
Memoir, 1863, 97
Milk, its Nature, 105
Modern Dairy Farming, 1916, 63

Modern Fruit Culture, 1892, 61
The Modern Householder, 100
Modern Husbandry, 1853, 5
Morton's Handbooks, 117

Natural Illustrations of British Grasses, 1846, 69
Natural Phenomena, 66
The New Testament of Agriculture, 1895, 120
The Noble Science, 113
Notebook of Agricultural Facts, 1883, 67
Notes on an Agricultural Tour in Belgium, Holland and the Rhine, 1862, 10
Notes and Descriptions, 1889, 93
Notes on Fields and Cattle, 1862, 74
Notes Historical and Practical on Farming and Farming Economy Vol II, 1863, 10
Notes on the Valuation, 1900, 117
Notes on the Warble Fly, 1884, 93

Observations on the Present Decay among Potatoes, 1845, 3
Odds and Ends, 74
On the More Frequent Growth of Barley, 1875, 42
Opening Lecture, 1878-9, 50
The Origin of Agriculture, 1886, 81
Original Chemistry, 58
Ornithology, 1893, 111
Our Farm of Four Acres, 1859, 9, 40, 74
Outlines of Farm Management, 1885, 106
Outlines of Farm Management and the Organisation of Farm Labour, 1880, 10
Outlines of Modern Farming, 1863,

INDEX OF TITLES

10
Pasture Grasses, 1887, 86
Pastures Old and New, 76
The Peasant's Home, 1876, 50
Permanent and Temporary, 1886, 83
The Perplexed Farmer, 1891, 101
Physiology on the Farm, 1867, 29
Pig Breeds, 1897, 124, 126
The Pig in Health, 1907, 129
Pigs for Profit, 1910, 91
Plain Letters, 58
Pomona Herefordensis, 1811, 80
The Poor, 1905, 190
Popular Botany, 1835, 3
Popular and Economic Botany, 1853, 23
The Potato and its Cultivation, 55
The Potato in Field and Garden, 1893, 103
Potatoes, or How to Grow, 1897, 54
Poultry Keeping, 1897, 57
Practical Dairy Farming, 1893, 57
The Practical Directory, 1881, 107
The Practical Directory for the Improvement of Landed Property, 1881, 10
Practical Farming, 1870, 37
Practical Forestry, 1888, 53
Practical Fruit Culture, 1892, 108
The Practical Guide, 92
A Practical Guide to Meat, 1909, 91
Practical Hints on Farming, 1864, 20
Practical Observations, 1889, 98
Practical Pig Keeping, 1892, 124
Practical Remarks on Agricultural Drainage, 1870, 40
Practical Water Farming, 1868, 32
Practice with Science, 1867-1869, 19

The Practice of Sheep Farming, 83
Principles of Agricultural Practice, 1888, 52, 109
Principles of Agriculture, 1827, 18, 73
Private Flock, 115
Profitable Clay Farming, 1881, 60
Profitable Dairy Farming, 1888, 93
Profitable Farming Handbooks, 90
Profitable Plants, 1865, 24
Purdon's Practical Farmer, 1863, 35

Rational Pig Keeping, 1893, 103
On the Reclamation and Protection of Agricultural Land, 1874, 45
Records of the Seasons, 1883, 66
The Relation of Chemistry, 1876, 48
On a Remarkable Effect of Cross Breeding, 1851, 51
Report on the Employment of, 88
Report on the Farming of Somersetshire, 1850, 17
Review of Agricultural Experiments, 1885, 18
The Revised English Agriculture, 109
Root Growing and Cultivation of Hops, 88
Rothamsted Experiments, 1888, 99, 123
Royal Agriculture, 1889, 96
Rudimentary Treatise on Agricultural Engineering, 1852-3, 6
Rural Economy, 1789 80
Rural England, 1901-2, 129
The Rural Exodus, 1892, 109
Rural Life Described, 1868-9, 33
Rust, Smut, Mildew, 1886, 79

Saddle and Sirloin, 92
The Science of Agriculture, 1854, 98
Secrets of Farming, 1863, 16
Sewage and its General Application 1869, 33
The Sewage Question, 1871, 25
Sewage Utilization, 1872, 39
Sheep Breeds, 1893, 126
Sheep Breeds and Management, 1908, 52
Sheep, Domestic Breeds, 1896, 80, 81
Sheep Farming, 1883, 67, 109
The Sheep and Lamb, 90
Sheep, their Management, 1871, 42
Sheep, Pigs, 1881, 114
Sheep, Pigs and Goats, 114
The Sheep and Pigs of Great Britain, 50
Sheep Raising, 104
The Sheep, its Varieties, 1873, 45
Short Notes on Silo Experiments, 1885, 68
Shorthorn Experiences, 1888, 91
The Shorthorn Herds, 1885, 92
Silos for Preserving British Fodder, 1883, 68
Six Centuries of Work and Wages, 1894, 68
The Skeleton at the Plough, 1874, 47
Small Farms, 1887, 90
The Soil of the Farm, 1884, 82
Soils and Manures, 104, 110
Soils and their Properties, 1890, 99
The Standard Encyclopaedia, 1908—11, 73
Stock Cattle, Sheep and Horses, 1863, 10
Stock Keeping and Cattle Rearing, 87

The Stockowners Guide, 65
Specimens of British Grasses, 1884, 69
On the Study of Botany, 1857, 23
Subject Investigations, 1879, 50
Successful Farming, 1870, 74
The Sugar Beet, 56
Sugar Beet, 1891, 103
Sulphate of Ammonia, 1886, 64
Sulphate of Ammonia, 1900, 72
Systematic Small Farming, 1886, 105

Text Book of Agricultural, 1892, 93
Text Book of Plants, 1899, 128
Theory and Practice, 1885, 77
Thorough Cultivation, 1895, 119
Tillage, 1891, 103
From Tolpuddle to TUC, 1948, 109
Tour through North America, 1835, 48
A Treatise on Agricultural Buildings,
A Treatise on Manures, 1889, 99
Tree Planting, 87
The True Art of Manuring, 27
The True Cause of Vine Disease, 1863, 16
The True Grasses, 1896, 61, 120
Two Letters on Cowkeeping, 1861, 15

Use of Salt in Agriculture, 1865, 26
The Utility of Sulphate, 1899, 117
The Utility of Town Sewage, 1863, 10

Valuation, 1891, 53
The Villa and Cottage Florists Directory, 1830, 3

Waste Products, 1876, 51
What Can Now Be Done for British Agriculture, 1842, 24

INDEX OF TITLES

Wheat. Its History, 1865, 25
The Wheat Problem, 1898, 40
White Cattle, 1879, 119
Wild White Cattle of Britain, 54

Women's Place, 1913, 73
Wool and its Applications, 1876, 24